Young Researcher Meeting in Rome 2012

Journal of Physics: Conference Series Volume 383

Rome, Italy
20 January 2012

Editors:

Fabio Agostini
Giordano Cattani
Luca Mazzaferro
Marina Migliaccio
Davide Pietrobon
Daniel Ricci Pacifici
Francesco Stellato
Marcella Veneziani

ISBN: 978-1-62276-850-9
ISSN: 1742-6588

Printed from e-media with permission by:

Curran Associates, Inc.
57 Morehouse Lane
Red Hook, NY 12571

Some format issues inherent in the e-media version may also appear in this print version.

Printed by Curran Associates, Inc. (2013)

For permission requests, please contact the Institute of Physics
at the address below.

Institute of Physics
Dirac House, Temple Back
Bristol BS1 6BE UK

Phone: 44 1 17 929 7481
Fax: 44 1 17 920 0979

techtracking@iop.org

Additional copies of this publication are available from:

Curran Associates, Inc.
57 Morehouse Lane
Red Hook, NY 12571 USA
Phone: 845-758-0400
Fax: 845-758-2634
Email: curran@proceedings.com
Web: www.proceedings.com

TABLE OF CONTENTS

Preface

At its third edition, the Young Researcher Meeting in Rome (YRMR) proves to be a growing event in the Italian scientific panorama. The high-quality content of the abstracts submitted to the scientific committee resulted in an exciting conference, held, for the second time, at the University of Rome 'Tor Vergata' on 20 January 2012.

A busy schedule covered a large variety of cutting-edge science topics: fundamental interactions, particle physics, cosmology, astrophysics, condensed matter and biomedical physics.

The broad range of the subjects discussed is the distinctive feature of the YRMR, a meeting aimed at enhancing the synergy among complementary branches of science by stimulating a fruitful exchange between theoretical, experimental and computational physics. Promoting collaborations between PhD students, postdoctoral fellows and young researchers creates a solid scientific network with an open-minded approach to discovery.

In this volume, we collect the contributions that have been presented both in the form of talks and of posters.

INFN UNIVERSITA' DEGLI STUDI DI ROMA TorVergata

Acknowledgments

The organizers of the 3rd Young Researcher Meeting in Rome would like to thank all the scientists who participated in the meeting. We thank the Science Faculty of the University of Rome 'Tor Vergata' for hosting the conference. The 'Tor Vergata' Division of the National Institute for Nuclear Physics (INFN) and the Department of Physics of the University of Rome 'Tor Vergata' who have sponsored the event, covering the organization costs and allowing for travel grants to the speakers. We are grateful to Professor Francesco Fucito, Professor Piergiorgio Picozza, Professor Rinaldo Santonico, Dr Jose Francisco Morales and Ms Francesca Luna for their valuable support.

Journal of Physics: Conference Series **383** (2012) 011001 doi:10.1088/1742-6596/383/1/011001

YRMR Organising and Editorial Committee

Fabio Agostini (fabio.agostini@roma2.infn.it)
Dipartimento di Fisica, Università di Roma 'Tor Vergata'
Via della Ricerca Scientifica 1, 00133 Roma
Italy
Giordano Cattani (giordano.cattani@roma2.infn.it)
Dipartimento di Fisica, Università di Roma 'Tor Vergata'
INFN sezione di Roma 'Tor Vergata'
Via della Ricerca Scientifica 1, 00133 Roma
Italy
Luca Mazzaferro (luca.mazzaferro@roma2.infn.it)
Dipartimento di Fisica, Università di Roma 'Tor Vergata'
INFN sezione di Roma 'Tor Vergata'
Via della Ricerca Scientifica 1, 00133 Roma
Italy
Marina Migliaccio (migliaccio@ifca.unican.es)
Instituto de Fisica de Cantabria, Edificio Juan Jorda,
Avenida de los Castros, E-39005 Santander, Cantabria
Spain
Davide Pietrobon (davide.pietrobon@jpl.nasa.gov)
Jet Propulsion Laboratory – California Institute of Technology
4800 Oak Grove Drive 169-237 91109 Pasadena, CA
USA
Daniel Ricci Pacifici (Daniel.Ricci.Pacifici@roma2.infn.it)
Dipartimento di Fisica, Università di Roma 'Tor Vergata'
INFN sezione di Roma 'Tor Vergata'
Via della Ricerca Scientifica 1, 00133 Roma
Italy
Francesco Stellato (francesco.stellato@cfel.de)
Center for Free Electron Laser science c/o DESY
22607 Hamburg
Germany
Marcella Veneziani (marcella.veneziani@caltech.edu)
California Institute of Technology
1200 California Blvd, Pasadena, 91125, CA
USA

Journal of Physics: Conference Series **383** (2012) 011002 doi:10.1088/1742-6596/383/1/011002

Peer review statement

All papers published in this volume of *Journal of Physics: Conference Series* have been peer reviewed through processes administered by the proceedings Editors. Reviews were conducted by expert referees to the professional and scientific standards expected of a proceedings journal published by IOP Publishing.

Vacua and Gaugings of Maximal $D = 4$ Supergravity

Gianluca Inverso

Dipartimento di Fisica, Università di Roma Tor Vergata
and I.N.F.N. - sezione di Roma 2, via della Ricerca Scientifica, 00133 Roma, Italy

E-mail: gianluca.inverso@roma2.infn.it

Abstract. I review an effective method to find vacua of gauged supergravity and apply it to the maximal theory in four dimensions. This leads in particular to new Anti de Sitter and Minkowski vacua with broken supersymmetry, as well as new gauge theories. Interesting links between different gaugings, their vacuum solutions and the stability properties as well as the allowed values of the cosmological constant are outlined.

1. Introduction

Maximal supergravity in 4 dimensions has received a lot of attention because of its unique matter content and its special properties. Despite the fact that the absence of chiral gauge interactions makes the theory less interesting for phenomenology, it is still at the centre of current investigations because of its possible perturbative finiteness and because its gaugings may be related via the AdS/CFT correspondence to the effective theories of stacks of M2-branes and hence find applications on strongly coupled gauge theories and condensed matter systems. Moreover, a systematic study of gauge maximal supergravity and its vacuum solutions deserves attention as it can help to understand why it is so difficult to build supersymmetric models giving rise to a vanishing and especially positive cosmological constant, and what ingredients in the compactification scheme may or may not help in such task.

Based on the results of the main paper [1], we discuss here how by using the embedding tensor formalism [2] for gauged supergravity and by taking advantage of the fact that the scalar manifold is a homogeneous space, a classification of the possible gauge models and their vacuum solutions in maximal supergravity can be simplified. Indeed, so far only a handful of analytical solutions has been found [3] and only recently numerical techniques have allowed for a more systematic scan of the vacua of the SO(8) theory [4]. Moreover, until now the only known way to break $\mathcal{N} = 8$ supergravity on a Minkowski vacuum has been the Scherk–Schwarz model or equivalently the reduction on twisted tori [5, 6], which lead to flat gauge groups of the type $\mathrm{U}(1) \ltimes T^r$. We will provide here some new examples.

Although a full classification of both gaugings and vacua of $\mathcal{N} = 8$ supergravity in four dimensions might still be a too ambitious goal and would probably require numerical computations, we will show how the methods applied here allow us to easily identify both new gaugings and interesting vacuum solutions.

2. Gauged maximal supergravity

The bosonic content of maximal $D = 4$ supergravity is given by the graviton, 28 vector fields A_μ^Λ and 70 scalars ϕ parametrizing the coset space $\mathrm{E}_{7(7)}/\mathrm{SU}(8)$. The ungauged theory can be seen as arising from the compactification of Type II or 11 dimensional supergravity on a torus. In that case the vectors are abelian gauge fields, but no matter is charged under the corresponding $U(1)^{28}$. $E_{7(7)}(\mathbb{R})$ acts on scalar fields as isometry transformations and on vectors as electromagnetic duality, which we recall is only a symmetry at the level of the equations of motion. In order to make this explicit, one can introduce "magnetic dual" vector fields $A_{\mu\,\Lambda}$ to form the **56** of $E_{7(7)}$: $A_\mu^M = (A_\mu^\Lambda,\, A_{\mu\,\Lambda})$. Of course only half of these vectors are physical fields. Gauged deformations of this theory are obtained when a subgroup of its global symmetries is promoted to local. This is most effectively achieved by means of the embedding tensor formalism [2], which allows to completely encode the choice of gauging in a single object and to treat all gaugings at the same footing in a formally $E_{7(7)}$-covariant fashion. We will denote the generators of $E_{7(7)}$ by t_α, $\alpha = 1, \ldots 133$. A subgroup G_{gauge} of $E_{7(7)}$ can then be gauged by introducing the embedding tensor $\Theta_M{}^\alpha$, so that for instance the gauge covariantization of derivatives is given by

$$D_\mu \equiv \partial_\mu - g A_\mu^M \Theta_M{}^\alpha t_\alpha, \qquad t_\alpha \in \mathfrak{e}_{7(7)}. \tag{1}$$

The matrix $\Theta_M{}^\alpha$ encodes how the algebra $\mathfrak{g}_{\text{gauge}}$ is embedded in $\mathfrak{e}_{7(7)}$, as well as what linear combinations of the vector fields A_μ^M are used to build the gauge connection. Note that also 'magnetic' vector fields $A_{\mu\Lambda}$ can take part in the gauge connection, and one can even gauge a group which is a symmetry only at the level of the equations of motion[1]. In any case, as only 28 vectors are physical, the rank of $\Theta_M{}^\alpha$ and hence the dimension of G_{gauge} can be at most 28. The embedding tensor must satisfy a set of algebraic constraints in order to define a consistent gauging. We define gauge generators $X_M \equiv \Theta_M{}^\alpha t_\alpha$ and will use in particular those in the **56** representation of $E_{7(7)}$: $X_{MN}{}^P \equiv \Theta_M{}^\alpha [t_\alpha]_N{}^P$. The constraints are:

$$\begin{aligned} \Theta_M{}^\alpha \Theta_N{}^\beta \Omega^{MN} &= 0, \qquad & [X_M, X_N] &= -X_{MN}{}^P X_P, \\ X_{MN}{}^M &= 0, \qquad & (t^\beta t_\alpha)_M{}^N \Theta_N{}^\alpha &= -\tfrac{1}{2}\Theta_M{}^\beta. \end{aligned} \tag{2}$$

The quadratic constraints guarantee locality of the theory and closure of the gauge algebra, while the linear ones are imposed by supersymmetry and force the embedding tensor to belong to the **912** $\subset$ **56** $\times$ **133** representation of $E_{7(7)}$.

3. Finding vacua of gauged maximal supergravity

The scalar potential that arises in gauged maximal supergravity is of course also parametrized by the embedding tensor. We denote the representatives of the coset space $E_{7(7)}/SU(8)$ by $L(\phi)_M{}^N$, which depend on the 70 scalar fields ϕ. We can write the scalar potential as:

$$V(\phi) = \frac{g^2}{672}\left(X_{MP}{}^R X_{NQ}{}^S \mathcal{M}(\phi)^{MN} \mathcal{M}(\phi)^{PQ} \mathcal{M}(\phi)_{RS} + 7 X_{MP}{}^R X_{NR}{}^P \mathcal{M}(\phi)^{MN}\right), \tag{3}$$

where $\mathcal{M}_{MN} \equiv (LL^T)_{MN}$ and $\mathcal{M}^{MN}$ is its inverse. Note in particular that $V(\phi)$ and $\mathcal{M}(\phi)$ are invariant under the local SU(8) transformations that define the coset. Studying this form of the scalar potential is very difficult, given that the dependence on scalar fields is at best given by a combination of exponentials and polynomials in 70 variables. Moreover one would have to study $V(\phi)$ separately for each possible gauging. We observe instead that the scalar potential only depends on the contraction of the embedding tensor with the coset representatives: $V(\phi, \Theta) = V(L(\phi)^{-1}\Theta)$.[2] As the scalar manifold is a homogeneous space, we can map any point (and hence any extremum of the potential) to the origin $\phi = 0$ by some $E_{7(7)}$ element U, provided

[1] however there always exist symplectic frames in which G_{gauge} is a symmetry of the action and magnetic fields do not appear in the gauge connection, namely $\Theta^{\alpha\Lambda} = 0$.

[2] we write $L(\phi)^{-1}\Theta$ as a shorthand notation for $L(\phi)^{-1}{}_M{}^N \Theta_N{}^\beta L(\phi)^{-1}{}_\beta{}^\alpha$

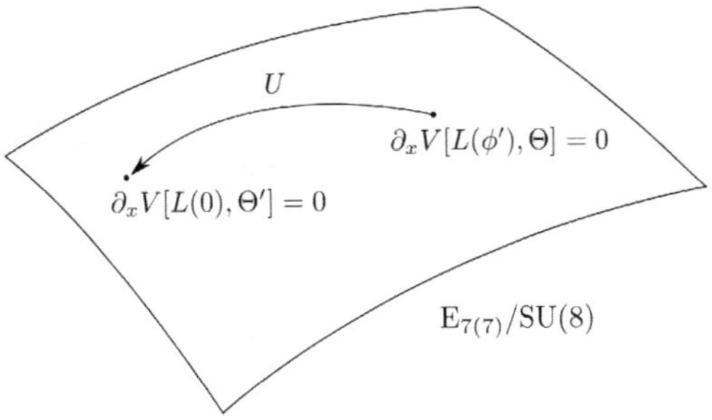

Figure 1. Stationary points of the scalar potential $V(\phi)$ found for $\phi = \phi'$ can always be moved back to $\phi = 0$ by means of an isometry $U \in E_{7(7)}$

that we redefine Θ by the action of the same transformation.[3] Therefore, given an appropriate U we can always write

$$V(\phi,\Theta) = V(0, U^{-1}\Theta), \quad U \in E_{7(7)}. \tag{4}$$

At such point, the scalar potential is a simple quadratic function of the embedding tensor and the problem of finding extrema is reduced to quadratic conditions on Θ [1, 7]. We have therefore translated the search for extrema of $V(\phi)$ into the study of the dependence of $V(\Theta) \equiv V(\phi = 0, \Theta)$ on the $E_{7(7)}$ orbits of the embedding tensor. Moreover, it is also possible to study the vacua of whole classes of gaugings together, by constructing embedding tensors that satisfy the constraints (2) as well as the stationary point conditions of $V(\Theta)$.

4. Gauge groups in $SL(8,\mathbb{R})$

We choose to consider the class of gaugings contained in $SL(8,\mathbb{R})$, which can be taken as the global symmetry of the action by an appropriate choice of symplectic frame. Under SL(8) the embedding tensor (which we recall sits in the **912** of $E_{7(7)}$) decomposes in the irreducible representations: $\mathbf{912} \to \mathbf{36} + \mathbf{36}' + \mathbf{420} + \mathbf{420}'$. The **36** and **36**$'$ in particular can be represented by symmetric matrices ξ^{AB} and θ_{AB} respectively, with $A, B, \ldots = 1, \ldots 8$. The corresponding form of the embedding tensor is:

$$\Theta_{AB,}{}^{C}{}_{D} \propto \delta^{C}_{[A}\theta_{B]D}, \qquad \Theta^{AB,C}{}_{D} \propto \delta^{[A}_{D}\xi^{B]C}. \tag{5}$$

The quadratic constraints when only these terms of the embedding tensor are switched on reduce to the conditions:

$$\xi = c\theta^{-1}, \; c \in \mathbb{R}, \quad \text{or} \quad \theta\xi = 0. \tag{6}$$

When $\xi = 0$ one recovers the well-know class of $SO(p,q)$ and $CSO(p,q,r)$ gaugings of maximal $D = 4$ supergravity [2, 8], with $p + q + r = 8$. They are classified by the positive, negative and vanishing eigenvalues of θ_{AB}, which plays the role of the quadratic invariant of the gauge algebra in the vector representation. In particular, $\theta = \mathbb{1}_8$ gives rise to the SO(8) gauged supergravity. The $CSO(p,q,r)$ groups are contractions of the semi-simple ones and arise when θ is singular. These gaugings are electric in the symplectic frame of SL(8) and also only involve electric vectors $A^{\Lambda}_{\mu} \equiv A^{[AB]}_{\mu}$. Turning on ξ, we can construct a different form of the $SO(p,q)$ gaugings and also obtain new non semisimple contractions when $\theta\,\xi = 0$:

$$G_{\text{gauge}} = \big(SO(p,q) \times SO(p',q')\big) \ltimes T^{(8-r)r+(8-r')(r'+r-8)}, \tag{7}$$

[3] the gauge algebra generated by $U^{-1}\Theta$ is generally only isomorphic to the original one and the vectors involved in the gauge connection may change as a consequence of the duality transformation.

where $p+q+r = p'+q'+r' = 8$. These are indeed new gaugings of maximal supergravity. They are still contained in SL(8), however now the gauge connection must involve magnetic vector potentials $A_{\mu\Lambda} = A_{\mu\,[AB]}$. A more detailed discussion of the semi-simple case $\xi \propto \theta^{-1}$ is given in [9].

As explained in the previous section, we can look for vacua of these classes of gauge theories by keeping the scalar fields to zero and studying the orbits of the embedding tensor under $E_{7(7)}$. However, the full orbits will necessarily involve the introduction of the **420** and **420′** components of the embedding tensor. We then restrict to studying the submanifold given by the coset space SL(8)/SO(8), generated by the real parts of the scalar fields. Even if this analysis is not going to be exhaustive, it still allows us to find new interesting results. Taking $\phi = 0$ in (3), a direct calculation of the potential gives

$$V(\theta', \xi') = \tfrac{1}{8}\mathrm{Tr}\,(\theta'^2) - \tfrac{1}{16}(\mathrm{Tr}\,\theta')^2 + \tfrac{1}{8}\mathrm{Tr}\,(\xi'^2) - \tfrac{1}{16}(\mathrm{Tr}\,\xi')^2, \tag{8}$$

and the stationary point condition can be cast in the form:

$$2(\theta^2 - \xi^2) - (\theta\mathrm{Tr}\theta - \xi\mathrm{Tr}\xi) \propto \mathbb{1}_8. \tag{9}$$

In other words, the traceless part of $2(\theta^2 - \xi^2) - (\theta\mathrm{Tr}\theta - \xi\mathrm{Tr}\xi)$ must vanish. The problem of finding vacua is now formulated in terms of the eigenvalues of θ and ξ. Although a full scan of the resulting vacuum solutions will likely still require numerics, we find a set of new analytic solutions, summarized in Table 1, where the spectra for all scalar fields are also included. All anti de Sitter vacua turn out to be unstable as they do not satisfy the Breitenlohner–Freedman bound. Minkowski vacua break supersymmetry completely. We refer to the main paper for a more thorough discussion of the results and presentation of all the explicit solutions. By performing a suitable symplectic transformation, one can show that the SO(8) vacua included here were known [3] and belong to the standard SO(8) theory, where the fields taking a vev are pseudoscalars, which also parametrize a (different) SL(8)/SO(8) coset space.

#	G_{gauge}	G_{res}	Λ_{cosm}	$m^{2\ (multipl.)}$
vi	SO(2, 6)	SO(2) × SO(6)	Mink	$2^{(2)}, \frac{1}{2}^{(20)}, 0^{(48)}$
	CSO(2, 0, 6)	SO(2)		
xii	$SO(4) \times SO(2,2) \ltimes T^{16}$	$SO(2)^2 \times SO(4)$	Mink	$4^{(4)}, 2^{(12)}, 1^{(16)}, 0^{(38)}$
xiii	$SO(2)^2 \ltimes T^{20}$	$SO(2)^2$		
i	SO(8)			
iii	SO(7, 1)	SO(7)	AdS	$2^{(1)}, -\frac{4}{5}^{(27)}, -\frac{2}{5}^{(35)}, 0^{(7)}$
viii	$SO(7) \ltimes T^7$			
ii	SO(8)			
iv	SO(7, 1)	SO(6)	AdS	$2^{(2)}, -1^{(20)}, -\frac{1}{4}^{(20)}, 0^{(28)}$
ix	$SO(7) \ltimes T^7$			
x	$SO(6) \times SO(1,1) \ltimes T^{12}$			
v	SO(7, 1)	SO(5)	AdS	$2^{(3)}, -\frac{4}{3}^{(14)}, \frac{2}{3}^{(5)}, 0^{(48)}$
xi	$SO(6) \times SO(1,1) \ltimes T^{12}$			
vii	SO(3, 5)	SO(3) × SO(5)	dS	$-2^{(1)}, 4^{(5)}, 2^{(30)}, \frac{4}{3}^{(14)}, -\frac{2}{3}^{(5)}, 0^{(15)}$

Table 1. mass spectra and residual symmetries for the new vacua. Known solutions of CSO(2, 0, 6) and SO(8) theories are given for reference. When $\Lambda \neq 0$, masses are normalized with respect to it, while we set $g = 1$. Supersymmetry is completely broken in all vacua.

One important thing we notice from Table 1 is that vacua belonging to different theories seem to share the same properties, at least up to the quadratic order in the expansion around the vacuum solution. Indeed, one should expect that multiplicities in the mass spectrum fall into representations of the residual gauge symmetry group. However, even the sign of the cosmological constant and the values of the masses turn out to coincide (for AdS vacua we actually refer to the ratio of the masses over the cosmological constant). For instance the SO(6, 2) shares the same spectrum of the CSO(2, 0, 6) $\simeq$ U(1) $\ltimes T^{12}$ theory, which is the Scherk–Schwarz model with 4 identical parameters [5]. This is also true for vector and fermion spectra. However beyond the quadratic level the two models are not identical, as moving around the moduli space of the SO(6, 2) solution one finds instabilities, which are absent in the SS models. Moreover, moving along some flat directions in the SO(6, 2) theory till the boundary of the moduli space allows us to recover vacua which reflect the gauge structure of another model. To give an example and in order to show the explicit form of at least some of the new vacua, we can parametrize two flat directions of the SO(6, 2) vacuum as:

$$\theta = \xi^{-1} = \text{diag}(-s,\ -s,\ s,\ s,\ u/s,\ u/s,\ 1/us,\ 1/us), \qquad s, u \geq 0. \tag{10}$$

By taking the singular limits $g \sim s \to 0$, $g \sim u \to 0$ and $g \sim s$, $s \sim u \to 0$ one recovers the other three Minkowski vacua in the table. More properties of the mass spectra of these vacua are discussed in [10].

As a last, more general, observation, we note that the scalar potential (3) taken at $\phi = 0$, in a symplectic frame where $\mathcal{M}(0) = \mathbb{1}_{56}$ and choosing orthonormal generators t_α, can be written as

$$V(\Theta) = \tfrac{g^2}{672}\Theta_M{}^\alpha \Theta_M{}^\beta \left(\delta^\alpha_\beta + 7\eta_{\alpha\beta}\right), \tag{11}$$

where $\eta_{\alpha\beta}$ is the Killing–Cartan metric on $\mathfrak{e}_{7(7)}$, with negative eigenvalues for compact generators. $V(\Theta)$ is clearly *not* $\text{E}_{7(7)}$ invariant. One sees that whenever the vectors of the gauge connection all have the same normalization, the counting of compact and non-compact generators in $\mathfrak{g}_{\text{gauge}}$ suffices to identify the sign of the cosmological constant. Although this argument is spoiled when acting with a general $\text{E}_{7(7)}$ transformation on Θ (and indeed, for example, the SO(8) potential is not negative definite), we observe that all the vacua known so far respect the pattern predicted by this simple observation.

Acknowledgements
I would like to thank the organizers of the Young Researchers Meeting 2012 for the opportunity to present this work. I thank G. Dall'Agata for collaboration on this project, and M. Bianchi, F. Catino, A. Marrani, F. Zwirner for helpful discussions.

References

[1] G. Dall'Agata and G. Inverso, Nucl. Phys. B **859** (2012) 70 [arXiv:1112.3345 [hep-th]].
[2] B. de Wit, H. Samtleben, M. Trigiante, Nucl. Phys. **B655** (2003) 93-126. [hep-th/0212239]. JHEP **0706** (2007) 049. [arXiv:0705.2101 [hep-th]].
[3] B. de Wit, H. Nicolai, Nucl. Phys. **B231** (1984) 506. N. P. Warner, Phys. Lett. **B128** (1983) 169. Nucl. Phys. **B231** (1984) 250. C. M. Hull, Class. Quant. Grav. **2** (1985) 343. C. M. Hull, N. P. Warner, Nucl. Phys. **B253** (1985) 650. Nucl. Phys. **B253** (1985) 675.
[4] T. Fischbacher, K. Pilch, N. P. Warner, [arXiv:1010.4910 [hep-th]]. T. Fischbacher, JHEP **1009** (2010) 068. [arXiv:0912.1636 [hep-th]]. [arXiv:1007.0600 [hep-th]]. [arXiv:1109.1424 [hep-th]].
[5] E. Cremmer, J. Scherk, J. H. Schwarz, Phys. Lett. **B84** (1979) 83.
[6] L. Andrianopoli, R. D'Auria, S. Ferrara, M. A. Lledo, JHEP **0207** (2002) 010. [hep-th/0203206].
[7] G. Dibitetto, A. Guarino, D. Roest, JHEP **1103** (2011) 137. [arXiv:1102.0239 [hep-th]].
[8] R. D'Auria, S. Ferrara, M. Trigiante, JHEP **0601** (2006) 081. [hep-th/0511158]. F. Cordaro, P. Fre, L. Gualtieri, P. Termonia and M. Trigiante, Nucl. Phys. B **532** (1998) 245 [hep-th/9804056]. G. Dall'Agata, N. Prezas, H. Samtleben, M. Trigiante, Nucl. Phys. **B799** (2008) 80-109. [arXiv:0712.1026 [hep-th]].
[9] G. Dall'Agata, G. Inverso, F. Trigiante, to appear soon.
[10] G. Dall'Agata, F. Zwirner [arXiv:1205.4711 [hep-th]].

Anomalous $U(1)$ and asymmetry

Andrea Mammarella

Physics Department, University of Rome, Tor Vergata, Via della Ricerca Scientifica 1, 00133, Rome, Italy

E-mail: andrea.mammarella@roma2.infn.it

Abstract. In the past years many possible extensions of the Standard Model (SM) have been investigated. If one of this model is revealed at the LHC, we will need tools to distinguish it among the many others studied. One possibility to achieve this goal is to utilize the forward-backward asymmetry. In this paper we calculate the asymmetry for a model in which there is an extra $U(1)$ anomalous symmetry. Furthermore, we show that the asymmetry can be used to impose constraints on the free parameters of the model.

1. Introduction

String theory results suggest that extra $U(1)$ models [1] can be strong candidates to extend the Standard Model (SM). This led us to develop the MiAUMSSM [2], in which the anomalies related to the extra $U(1)$ are cancelled by the Green-Schwarz (GS) mechanism, thus letting the new charges unconstrained. An alternative version of this model which admits spontaneous supersymmetry breaking can be found in [3].
We have already studied the phenomelogy of this model using cosmological constraints, such as WMAP data on Dark Matter [4, 5]. We have also studied the radiative decays of the Next to Lightest Supersymmetric Particle in this model [6, 7].
In this paper we aim to use the asymmetry to investigate the detailed properties of the Z', the new gauge boson associated to the extra $U(1)$ by fixing the charges. We also develop a method that, given the experimental data, can confirm or discard the model in consideration. This method is based on the use of different definitions of asymmetry at the LHC.
At the LHC [8] cuts on the parameter space have to be performed. Each possible cut leads to a different definition of asymmetry at the LHC. In the paper we choose some of these definitions to make our calculations. We will add further details on their explicit form in the related section.
The article is organized as follow: in sec. 2 we briefly review the properties of the model that we are going to use, in sec. 3 we discuss the four definitions of asymmetry that will appear in the following and the related calculations. In sec. 4 we show our results.

2. MiAUMSSM

Our model [2] is an extension of the MSSM with an extra $U(1)$ gauge symmetry. The charges of the matter fields with respect to the symmetry gauge groups are given in table 1.
Gauge invariance of the model implies that only three of the $U(1)$ charges are independent. We have chosen Q_Q, Q_L and Q_{H_u} without losing generality. The anomalies induced by this extension are cancelled by the GS mechanism, so there are not further constraints on the charges.
We are going to evaluate the asymmetry of the process $pp \to e^+e^-$.

	$SU(3)_c$	$SU(2)_L$	$U(1)_Y$	$U(1)'$
Q_i	**3**	**2**	1/6	Q_Q
U_i^c	**3**	**1**	−2/3	Q_{U^c}
D_i^c	**3**	**1**	1/3	Q_{D^c}
L_i	**1**	**2**	−1/2	Q_L
E_i^c	**1**	**1**	1	Q_{E^c}
H_u	**1**	**2**	1/2	Q_{H_u}
H_d	**1**	**2**	−1/2	Q_{H_d}

Table 1. Charge assignment.

3. Four asymmetries at the LHC

The initial LHC state (pp) is symmetric. This implies that the total asymmetry at the LHC is zero if we integrate over the whole parameter space. However the partonic subprocess $q\bar{q} \to e^+e^-$ is asymmetric. To keep this asymmetry we have to impose kinematical cuts. Each possible way to perform these cuts leads to a different definition of asymmetry at the LHC.
In this work we use the four definitions of asymmetry described in [9]:

$$A_{\mathrm{RFB}}(Y^{\mathrm{cut}}_{e\bar{e}}) = \left.\frac{\sigma(|Y_e| > |Y_{\bar{e}}|) - \sigma(|Y_e| < |Y_{\bar{e}}|)}{\sigma(|Y_e| > |Y_{\bar{e}}|) + \sigma(|Y_e| < |Y_{\bar{e}}|)}\right|_{|Y_{e\bar{e}}|>Y^{cut}_{e\bar{e}}} \tag{1}$$

$$A_{\mathrm{OFB}}(p^{\mathrm{cut}}_{Z,e\bar{e}}) = \left.\frac{\sigma(|Y_e| > |Y_{\bar{e}}|) - \sigma(|Y_e| < |Y_{\bar{e}}|)}{\sigma(|Y_e| > |Y_{\bar{e}}|) + \sigma(|Y_e| < |Y_{\bar{e}}|)}\right|_{|p_{z,e\bar{e}}|>p^{\mathrm{cut}}_{Z,e\bar{e}}}, \tag{2}$$

$$A_{\mathrm{C}}(Y_{\mathrm{C}}) = \frac{\sigma_e(|Y_e| < Y_{\mathrm{C}}) - \sigma_{\bar{e}}(|Y_{\bar{e}}| < Y_{\mathrm{C}})}{\sigma_e(|Y_e| < Y_{\mathrm{C}}) + \sigma_{\bar{e}}(|Y_{\bar{e}}| < Y_{\mathrm{C}})} \tag{3}$$

$$A_{\mathrm{E}}(Y_{\mathrm{C}}) = \frac{\sigma_e(Y_{\mathrm{C}} < |Y_e|) - \sigma_{\bar{e}}(Y_{\mathrm{C}} < |Y_{\bar{e}}|)}{\sigma_e(Y_{\mathrm{C}} < |Y_e|) + \sigma_{\bar{e}}(Y_{\mathrm{C}} < |Y_{\bar{e}}|)} \tag{4}$$

The first two asymmetries are defined in the center of mass frame. The forward-backward asymmetry A_{RFB} [1, 10, 11, 12, 13, 14] has a cut on the rapidity Y of the $f/\bar{f}$ pair (in this work f is an electron), while the one side asymmetry A_O [15, 16] has a cut on $p_{z,f\bar{f}}$, the longitudinal component of the fermionic pair momentum. The other two asymmetries are defined in the lab frame. The variable Y_C is the pseudo-rapidity $-\log(tan(\theta/2))$: in this case the kinematical cut is on the angles of the outgoing particles with respect to the beam direction in the lab frame. The central asymmetry A_C [17, 18, 19, 20, 21] is calculated integrating in the angular region centered on the axis othogonal to the beam, while the edge asymmetry A_E [22] is defined in the complementar region.

4. Calculations and results

We calculate the value of the cuts that maximizes the significance of each definition of asymmetry, then we use these best cuts to evaluate the asymmetries in different ways, i.e. as a functions of two free charges keeping the third fixed to 0 and as a function of the three charges. We have set the mass of the Z' boson at 1.5 TeV to permit a direct comparison with the results of [9].

4.1. Optimal cuts

As shown in [9] the asymmetry magnitude is not a good function to optimize the observable. A better choice is, instead, the statistical significance:

$$Sig \equiv A\sqrt{\mathcal{L}\sigma} \tag{5}$$

where A can be any of the previously defined asymmetries, $\mathcal{L}$ is the LHC integrated luminosity, which we take to be 100 fb^{-1} and σ is the total cross section.
In figure 1 we show the results associated to the significance for all the asymmetry definition that we use in the "on peak "region, that is $M_{Z'} - 3\Gamma_{Z'} < M_{e^+e^-} < M_{Z'} + 3\Gamma_{Z'}$.
The best cuts are those that maximize the significance. We list them in the table 2

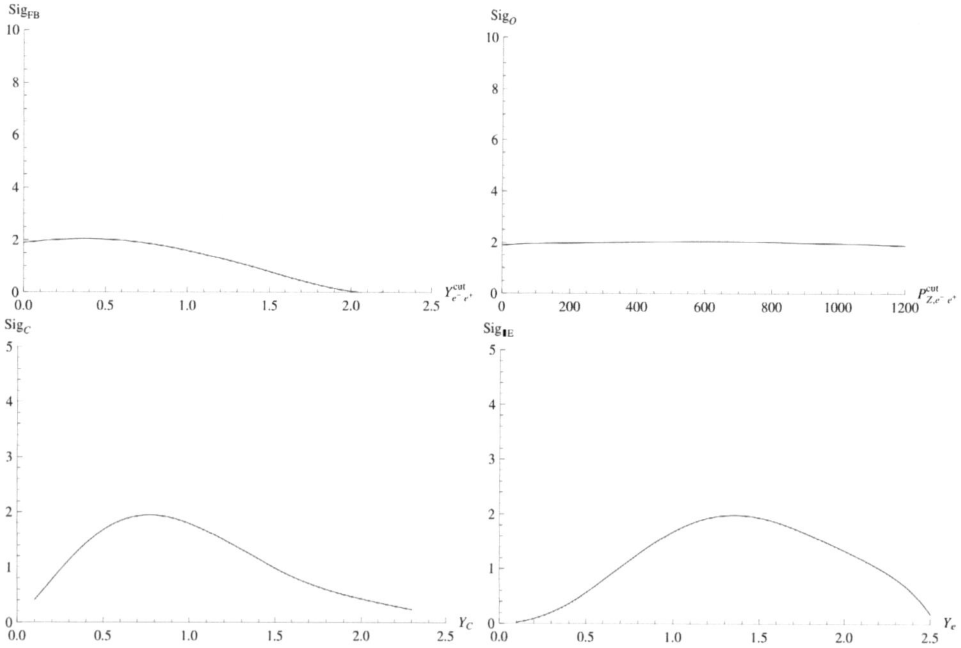

Figure 1. Significance of the four definitions of asymmetry as a function of the corresponding cut for for on-peak events

	A_{RFB}	A_O	A_C	A_E
Best cut	$Y^{cut}_{f\bar{f}} = 0.4$	$p^{cut}_{z,f\bar{f}} = 580\ GeV$	$Y_C = 0.8$	$Y_C = 1.4$

Table 2. Best cuts for the on-peak e^+e^- asymmetries

4.2. Asymmetry calculations

Using the best cuts we have studied the dependence of the asymmetry from couples of the free charges. There are three possibilities for the couples of the charges: $Q_{H_u} - Q_L$ ($Q_Q = 0$), $Q_{H_u} - Q_Q$ ($Q_L = 0$) and $Q_L - Q_Q$ ($Q_{H_u} = 0$). For the sake of brevity, in figures 2 and 3 we show only the plots obtained keeping $Q_L = 0$.

4.3. Charges constraints

To impose constraints on the charges, we have calculated the asymmetry dependence from the three charges. It turns out that the asymmetries are rational functions of 4th order in the

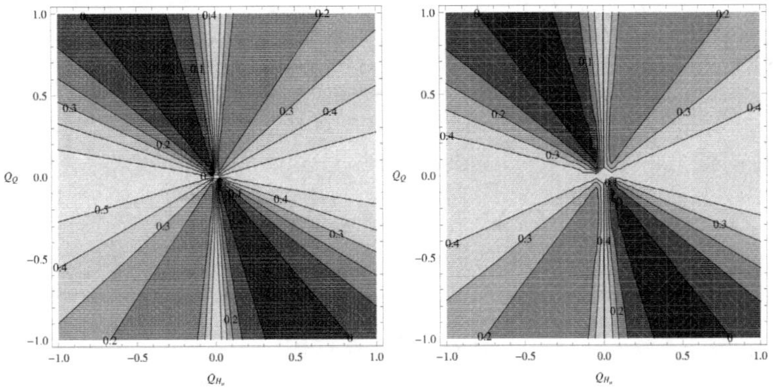

Figure 2. Results with the asymmetries for $Q_L = 0$ for the best cuts. The left image is A_{RFB}, the right image is A_O

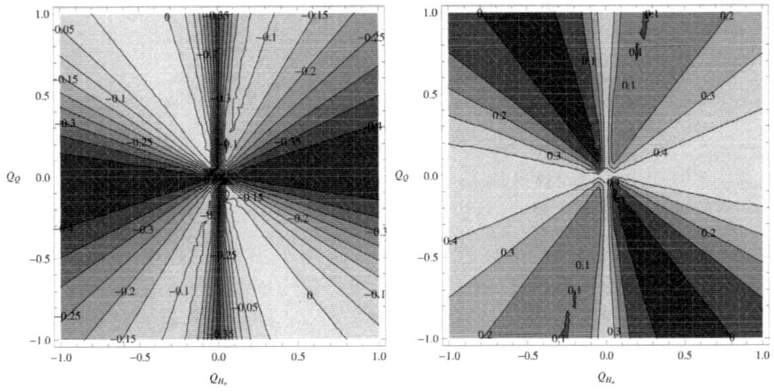

Figure 3. Results for the asymmetries with $Q_L = 0$ for the best cuts. The left image is A_C, the right image is A_E

charges, so we use the following fit function:

$$A = \frac{\sum_{i,j,k=0}^{n} a_{ijk}(Q_{H_u})^i (Q_Q)^j (Q_L)^k}{\sum_{i,j,k=0}^{n} b_{ijk}(Q_{H_u})^i (Q_Q)^j (Q_L)^k} \tag{6}$$

with $i + j + k = n \leq 4$. We have calculated this expansion for the previous four asymmetry definitions that we have considered so we have them expressed as a functions of the charges. We do not show here their complete expression, that can be found in However, if a deviation from the asymmetry associated to the SM asymmetry is discovered, this expressions can be used as a system of four equation of the three variables Q_{H_u}, Q_L and Q_Q. Solving this system using the experimental values of the asymmetry will give constraints on the charges and a consistency check of the model.

5. Conclusion

We have studied four different definitions of asymmetry at the LHC, namely the forward-backward, the one-side, the central and the edge asymmetries in the channel $pp \to e^+e^-$.
To perform this study we have written a numerical code that finds the value of the cut that maximizes the significance of each definition of asymmetry. After that, we have calculated the value of the four asymmetries as functions of couple of charges, keeping one of the three free

charges of our model fixed to 0.
Finally, we have calculated the functional dependence of the asymmetries from the three charges using an expansion in rational functions. If an asymmetry of this type will be found at the LHC this expansions could give constraints on the free charges of the model.
So we have completely characterized the asymmetry of the MiAUMSSM with a procedure that can be replied for every other model of physics beyond the SM.

References

[1] P. Langacker, "The Physics of Heavy Z' Gauge Bosons", Rev. Mod. Phys. **81** (2009) 1199 [arXiv:0801.1345 [hep-ph]].
[2] P. Anastasopoulos, F. Fucito, A. Lionetto, G. Pradisi, A. Racioppi and Y. S. Stanev, "Minimal Anomalous U(1)-prime Extension of the MSSM", Phys. Rev. D **78** (2008) 085014 [arXiv:0804.1156 [hep-th]].
[3] A. Lionetto and A. Racioppi, "Supersymmetry Breaking in a Minimal Anomalous Extension of the MSSM", arXiv:1102.5040 [hep-ph].
[4] F. Fucito, A. Lionetto, A. Mammarella and A. Racioppi, "Stueckelino dark matter in anomalous U(1)-prime models", Eur. Phys. J. C **69** (2010) 455 [arXiv:0811.1953 [hep-ph]].
[5] F. Fucito, A. Lionetto and A. Mammarella, Phys. Rev. D **84** (2011) 051702 [arXiv:1105.4753 [hep-ph]].
[6] A. Lionetto and A. Racioppi, "Gaugino radiative decay in an anomalous U(1)-prime model", Nucl. Phys. B **831** (2010) 329 [arXiv:0905.4607 [hep-ph]].
[7] F. Fucito, A. Lionetto, A. Racioppi and D. Ricci Pacifici, "A Phenomenological study on the wino radiative decay in anomalous $U(1)'$ models", Phys. Rev. D **82** (2010) 115004 [arXiv:1007.5443 [hep-ph]].
[8] E. Lyndon and B. Philip (ed.), *LHC Machine*, JINST **3** (2008) S08001.
[9] Z. q. Zhou, B. Xiao, Y. k. Wang and S. h. Zhu, "Discriminating Different Z's via Asymmetries at the LHC", Phys. Rev. D **83** (2011) 094022 [arXiv:1102.1044 [hep-ph]].
[10] P. Langacker, R. W. Robinett, J. L. Rosner, "New Heavy Gauge Bosons in p p and p anti-p Collisions", Phys. Rev. **D30** (1984) 1470.
[11] F. Petriello and S. Quackenbush, "Measuring Z' couplings at the CERN LHC", Phys. Rev. D **77** (2008) 115004 [arXiv:0801.4389 [hep-ph]].
[12] M. Cvetic and S. Godfrey, "Discovery and identification of extra gauge bosons", arXiv:hep-ph/9504216.
[13] M. Dittmar, A. S. Nicollerat and A. Djouadi, "Z-prime studies at the LHC: An Update", Phys. Lett. B **583** (2004) 111 [arXiv:hep-ph/0307020].
[14] S. Godfrey, T. A. W. Martin, "Identification of Extra Neutral Gauge Bosons at the LHC Using b- and t-Quarks", Phys. Rev. Lett. **101**, 151803 (2008). [arXiv:0807.1080 [hep-ph]].
[15] Y. -k. Wang, B. Xiao, S. -h. Zhu, "One-side forward-backward asymmetry at the LHC", Phys. Rev. **D83** (2011) 015002. [arXiv:1011.1428 [hep-ph]].
[16] Y. -k. Wang, B. Xiao, S. -h. Zhu, Phys. Rev. **D82** (2010) 094011. [arXiv:1008.2685 [hep-ph]].
[17] P. Ferrario and G. Rodrigo, "Charge asymmetries of top quarks: A Window to new physics at hadron colliders", J. Phys. Conf. Ser. **171** (2009) 012091 [arXiv:0907.0096 [hep-ph]].
[18] J. H. Kuhn, G. Rodrigo, "Charge asymmetry in hadroproduction of heavy quarks", Phys. Rev. Lett. **81** (1998) 49-52. [hep-ph/9802268].
[19] J. H. Kuhn, G. Rodrigo, "Charge asymmetry of heavy quarks at hadron colliders", Phys. Rev. **D59** (1999) 054017. [hep-ph/9807420].
[20] O. Antunano, J. H. Kuhn, G. Rodrigo, "Top quarks, axigluons and charge asymmetries at hadron colliders", Phys. Rev. **D77** (2008) 014003. [arXiv:0709.1652 [hep-ph]].
[21] P. Ferrario, G. Rodrigo, "Massive color-octet bosons and the charge asymmetries of top quarks at hadron colliders", Phys. Rev. **D78** (2008) 094018. [arXiv:0809.3354 [hep-ph]].
[22] B. Xiao, Y. K. Wang, Z. Q. Zhou and S. h. Zhu, "Edge Charge Asymmetry in Top Pair Production at the LHC", Phys. Rev. D **83** (2011) 057503 [arXiv:1101.2507 [hep-ph]].

Young Researcher Meeting in Rome 2012 IOP Publishing
Journal of Physics: Conference Series **383** (2012) 012003 doi:10.1088/1742-6596/383/1/012003

Exact solutions of Einstein and Einstein-Maxwell gravity minimally coupled to a scalar field

Matteo Serra

Dipartimento di Fisica, Università di Cagliari and INFN, Cittadella Universitaria, 09042 Monserrato, Italy

E-mail: matteo.serra@ca.infn.it

Abstract. We present a general method for solving exactly the static field equations of Einstein and Einstein-Maxwell gravity minimally coupled to a scalar field. Through this method one can derive broad classes of static solutions with radial symmetry of the theory, which may play an important role in applications of the AdS/CFT correspondence to condensed matter and strongly coupled QFT's. Moreover, the method allows to prove a new no-hair theorem about the existence of hairy black brane and black hole solutions.

1. Introduction

In recent years there has been a renewed growing interest for the static black hole solutions of Einstein (and Einstein-Maxwell) gravity coupled to scalar fields. In particular, the recent advances in string theory and the AdS/CFT correspondence [1] have shifted the focus from the search of asymptotically flat to asymptotically anti-de Sitter (AdS) solutions. In fact, static black hole and black brane solutions with a non trivial profile of the scalar field (scalar 'hair') and AdS asymptotics can play a crucial role in applications of the AdS/CFT correspondence to holographic strongly coupled quantum field theories (QFTs). In these theories the nontrivial, coordinate-dependent scalar hair of the black hole solutions can be interpreted as a scalar condensate which gives rise to a rich phenomenology in the dual QFT, reminiscent of well-known condensed matter systems [2, 3, 4, 5].

However, despite the growing importance played by static black hole solutions with scalar hair and AdS asymptotics, very few exact analytical solutions of this kind are known [6, 7, 8].

Starting from these considerations, we propose a general method for solving the field equations of Einstein and Einstein-Maxwell gravity minimally coupled to a scalar field ϕ in the static, radially symmetric case. This method allows to find exact analytic solutions of Einstein and Einstein-Maxwell gravity with scalar hair in several situations: four or generic $d+2$ spacetime dimensions, different topologies of the transverse space (planar, spherical, hyperbolic) and different asymptotics (anti-de Sitter, domain wall, conformal to Lifshitz spacetime). Moreover, the method can be used to prove a new 'no-hair' theorem about the general existence of black brane and black hole solutions of Einstein and Einstein-Maxwell gravity minimally coupled to a scalar field.

In this paper, mainly based on the work [9], we briefly present this solving method (Sect. 2), a panoramic of the exact solutions obtained (Sect.3, 4) and the formulation of the new no-hair theorem (Sect. 5). Finally, in Sect.6 we point out our concluding remarks.

2. The solving method

We start from the action of Einstein-Maxwell gravity in $d+2$ dimensions, minimally coupled to a scalar field ϕ and with a generic self-interaction potential $V(\phi)$:

$$A = \int d^{d+2}x\sqrt{-g}\left(\mathcal{R} - 2(\partial\phi)^2 - F^2 - V(\phi)\right). \tag{1}$$

We will investigate static, radially symmetric and purely electric solutions. We adopt a Schwarzschild gauge to write the spacetime metric:

$$ds^2 = -U(r)dt^2 + U^{-1}(r)dr^2 + R^2(r)d\Omega^2_{(\varepsilon,d)}, \tag{2}$$

where $\varepsilon = 0, 1, -1$ denotes, respectively, the d-dimensional planar, spherical, or hyperbolic transverse space with metric $d\Omega^2_{(\varepsilon,d)}$. In these coordinates, the electric field reads: $F_{tr} = Q/R^d$, where Q is the electric charge. With the parametrization (2), the field equations take the form:

$$\frac{R''}{R} = -\frac{2}{d}(\phi')^2, \quad (UR^d\phi')' = \frac{1}{4}R^d\frac{dV}{d\phi}, \tag{3}$$

$$(UR^d)'' = \varepsilon d(d-1)R^{d-2} + 2\frac{d-2}{d}\frac{Q^2}{R^d} - \frac{d+2}{d}R^dV, \tag{4}$$

$$(UR^{d-1}R')' = \varepsilon(d-1)R^{d-2} - \frac{2}{d}\frac{Q^2}{R^d} - \frac{1}{d}R^dV. \tag{5}$$

Our method for solving the field equations (3)-(5) works as follows. Assuming that the r-dependence of the scalar field $\phi(r)$ is given, and introducing the new variables F, Y and u defined as:

$$F(r) = -\frac{2}{d}(\phi')^2, \quad R = e^{\int Y}, \quad u = UR^d, \tag{6}$$

the field equations (3)-(5) become:

$$Y' + Y^2 = F, \quad (u\phi')' = \frac{1}{4}e^{d\int Y}\frac{dV}{d\phi}, \tag{7}$$

$$u'' - (d+2)(uY)' = -2\varepsilon(d-1)e^{(d-2)\int Y} + 4Q^2e^{-d\int Y}, \tag{8}$$

$$u'' = \varepsilon d(d-1)e^{(d-2)\int Y} + 2\frac{d-2}{d}Q^2e^{-d\int Y} - \frac{d+2}{d}e^{d\int Y}V. \tag{9}$$

The first equation in (7) is a first-order nonlinear equation for Y, known as the Riccati equation, which can be solved in a number of cases. Once the solution for Y has been found, we can integrate Eq. (8), which is linear in u, to obtain:

$$u = R^{d+2}\left[\int\left(4Q^2\int\frac{1}{R^d} - 2\varepsilon(d-1)\int R^{d-2} - C_1\right)\frac{1}{R^{d+2}} + C_2\right], \tag{10}$$

where C_1 and C_2 are integration constants. Finally, we can determine the potential $V(\phi)$ by using Eq. (9):

$$V = \frac{d^2(d-1)}{d+2}\frac{\varepsilon}{R^2} + 2\frac{d-2}{d+2}\frac{Q^2}{R^{2d}} - \frac{d}{d+2}\frac{u''}{R^d}, \tag{11}$$

while the metric functions read (cfr. 6): $R = \Lambda e^{\int Y}$, $U = u/R^d$, where the integration constant Λ coming from the integral of Y.

The key feature of this method is to assume a given profile $\phi(r)$ for the scalar field and to solve the system for the metric functions and the potential $V(\phi)$. This unusual approach is actually very useful because, focusing on solutions with AdS asymptotics, in particular for what concerns application to the AdS/CFT correspondence, the actual exact form of the potential $V(\phi)$ is not particularly relevant, while is often more important the behavior of the scalar field $\phi(r)$, in particular its fall-off behavior at $r = \infty$.

3. Uncharged solutions

In this section we will consider the case of (3 + 1)-dimensional spacetime, i.e. $d = 2$, and uncharged ($Q = 0$) black brane ($\varepsilon = 0$) solutions. In particular we are interested in asymptotically domain wall and asymptotically anti de-Sitter solutions.

3.1. Domain wall solutions

Our method for solving the field equations (3)-(5) requires an ansatz for the scalar field. Usually, domain wall solutions appear when the scalar behaves as $\log r$ [10, 11] . The most natural ansatz is therefore:

$$\gamma\phi = \log\frac{r}{r_-}, \tag{12}$$

where γ is a constant and r_- sets a length-scale.

With the ansatz (12), the Riccati equation in (7) is solved by:

$$Y = \frac{\alpha}{r}, \quad \alpha(\alpha-1) = -\frac{1}{\gamma^2}. \tag{13}$$

Parametrizing α and γ as: $\gamma^{-1} = h\alpha = h/(h^2+1)$, the solution takes the form:

$$U = \left(\frac{r}{r_-}\right)^{\frac{2}{1+h^2}} - C_1\left(\frac{r}{r_-}\right)^{-\frac{1-h^2}{1+h^2}}, \qquad R = \left(\frac{r}{r_-}\right)^{\frac{1}{1+h^2}}, \tag{14}$$

while the potential has a simple exponential form:

$$V = -\frac{2(3-h^2)}{(1+h^2)^2 r_-^2}\, e^{-2h\phi}. \tag{15}$$

This solution has the following main features:

- In the 'extremal' zero-temperature case, $C_1 = 0$, the solution (14) has the typical form of a single-scalar domain wall solution: $ds^2 = (Ar)^{\delta}(\eta_{\mu\nu}dx^\mu dx^\nu) + (Ar)^{-\delta}dr^2$, which preserves only the Poincaré isometry of the 3D transverse space [11];
- For $h^2 \leq 3$ and $C_1 > 0$, our solution (14) represents a black brane with domain wall asymptotics, a singularity at $r = 0$ and a regular horizon at $r_h = C_1^{(1+h^2)/(3-h^2)} r_-$.

3.2. Asymptotically AdS solutions

Inspired by known solutions in flat spacetime and in gauged supergravity [6], to search for asymptotically AdS solutions we use an ansatz in which ϕ is expressed in terms of an harmonic function X:

$$\gamma\phi = \log X, \qquad X = 1 - \frac{r_-}{r}, \tag{16}$$

where γ and r_- are constants.

With the ansatz (16), the Riccati equation (7) can be solved in terms of the harmonic function X to give:

$$R = \Lambda r X^{\beta+\frac{1}{2}}, \qquad \beta^2 - \frac{1}{4} = -\frac{1}{\gamma^2}. \tag{17}$$

We consider the $C_1 = 0$ extremal solution. The constant C_2 essentially determines the normalization of the potential and it can be fixed by choosing $C_2 = 1/L^2$. With these assumptions, Eq. (10) and (11) give, respectively, the solution for the metric and the scalar potential:

$$U = R^2 = \frac{r^2}{L^2}\left(1 - \frac{r_-}{r}\right)^{2\beta+1}, \tag{18}$$

$$V(\gamma,\phi) = -\frac{2}{L^2}e^{2\gamma\beta\phi}[2 - 8\beta^2 + (1 + 8\beta^2)\cosh(\gamma\phi) - 6\beta\sinh(\gamma\phi)]. \quad (19)$$

One can easily check that the hairy extremal solution (18) interpolates between an AdS vacuum: $ds^2 = (r/L)^2(\eta_{\mu\nu}dx^\mu dx^\nu) + (r/L)^{-2}dr^2$ at $r = \infty$ and a domain wall solution like the (14), with $C_1 = 0$, near $r = r_-$.

Unfortunately, our method does not allow to find a full $C_1 \neq 0$ one-parameter family of black brane solutions. If they exist, have to be found numerically.

4. Other solutions

We now present a brief overview on charged, spherical and $d+2$-dimensional solutions obtained through the solving method. For more details see [9].

4.1. Solutions conformal to Lifshitz spacetime

Starting again from the ansatz (12), we can obtain an exact solution of the field equations also when $Q \neq 0$. In this case, the potential has also a simple exponential form, while in the extremal $C_1 = 0$ case the solution is conformal to the Lifshitz spacetime: $ds^2 = l^2\left(-r^{2z}dt^2 + \frac{dr^2}{r^2} + r^2dx^idx^i\right)$, where z is the critical exponent [12].

For $C_1 > 0$ and in a specific range of the parameters, the solution represents a black brane with asymptotics conformal to Lifshitz, a singularity at $r = 0$ and a regular horizon.

4.2. Charged asymptotically AdS solutions

Solving the field equations with the ansatz (16), $Q \neq 0$ and $C_1 = 0$, we obtain a solution interpolating between an asymptotic AdS vacuum at $r = \infty$ and a conformal Lifshitz solution in the near-horizon limit $r \sim r_-$. This solution describes a one-parameter (the charge Q) family of black brane solutions, while also in this case the method does not allow to find a full $C_1 \neq 0$ two-parameter (mass and charge) family of solutions.

4.3. Other solutions

Several of the $\varepsilon = 0$ and $d = 2$ solutions found can be easily generalized to the case of spherical or hyperbolic topology $\varepsilon = \pm 1$ (black holes) and to $d + 2$ dimensions. In particular one can derive:

- Uncharged and charged black holes with Lifshitz and AdS asymptotics;
- Spherical solutions generated from the planar ones;
- $(d+2)$-dimensional domain wall, Lifshitz-like and uncharged asymptotically AdS solutions;

5. No-hair theorem

An important issue in this context is the question about the existence of regular, static black brane (hole) solutions of Einstein-scalar gravity with AdS asymptotics beyond the Schwarzschild and Reissner-Nordstrom solutions, i.e. of solutions endowed with non trivial scalar hair.

This problem has been already discussed in the literature. In particular, it has been argued that a necessary condition for the existence of such black hole solutions is the violation of the Positive Energy Theorem (PET) [13]. This result rules out black hole solutions with scalar hair and positive squared-mass m^2, but allows for the existence of these solutions when m^2 is negative but above the BF bound [14].

Through the solving method discussed it is possible to prove a new no-hair theorem about the existence of regular hairy black brane solutions of Einstein-scalar gravity, in which a key ingredient is the existence of an extremal $T = 0$ hairy black brane solution. The enunciate of our theorem is the following (for a detailed demonstration see [9]):

1) *One-parameter families of asymptotically AdS black brane solutions with nontrivial scalar hair exist only if the field equations (3)-(5) admit an extremal* $T = 0$, $U = R^2$ *solution.*
2) *Black-brane solutions that asymptotically approach the domain wall solution (14) exist in some range of the parameters for the case of an exponential potential* $V(\phi)$.
3) *The allowed asymptotically AdS hairy black brane solutions necessarily have a scalar hair that depends on the black brane temperature* T. *Solutions with temperature-independent scalar hair exist only for the case of domain wall spacetimes (14).*

It is important to stress that this no-hair theorem can be easily extended to the charged case $Q \neq 0$ and to the spherical (hyperbolic) case $\varepsilon = \pm 1$ (black holes).

6. Conclusions

We have presented a general method for finding exact, static and radially symmetric analytic solutions of Einstein and Einstein-Maxwell gravity minimally coupled to a scalar field. Rather than assuming a particular form of the scalar self-interaction potential, our method starts from an ansatz for the scalar field profile and determines, together with the metric functions, the corresponding form of the potential. For this reason it is particularly suitable for applications to the AdS/CFT correspondence. Through this method one can derive broad classes of exact analytic hairy solutions of Einstein and Einstein-Maxwell gravity minimally coupled to a scalar, and a new no-hair theorem about the existence of black brane (hole) solutions of Einstein gravity with scalar hair.

Our approach has a main drawback: in some situations it does not allow to find a full one-parameter family of black holes, i.e. the full spectrum of solutions for different temperatures, but only 'extremal' $T = 0$ solutions, and it is not even clear whether or not such solutions exist. A final answer to this question involves numerical computation, currently in progress.

Acknowledgments

M.S. gratefully acknowledges Sardinia Regional Government for the financial support of his PhD scholarship (P.O.R. Sardegna F.S.E. Operational Programme of the Autonomous Region of Sardinia, European Social Fund 2007-2013 - Axis IV Human Resources, Objective I.3, Line of Activity I.3.1).

References

[1] J M Maldacena 1998, *Adv. Theor. Math. Phys.* **2** 231
[2] S A Hartnoll, C P Herzog and G T Horowitz 2008, *JHEP* **1208** 015
[3] G T Horowitz and M M Roberts 2008, *Phys. Rev.* D **78** 126008
[4] M Cadoni, G D'Appollonio and P Pani 2010, *JHEP* **1003** 100
[5] M Cadoni and P Pani 2011, *JHEP* **1104** 049
[6] M J Duff and J T Liu 1999, *Nucl. Phys.* B **554** 237
[7] C Martinez, R Troncoso and J Zanelli 2004, *Phys. Rev.* D **70** 084035
[8] C J Gao and S N Zhang 2004, *Phys. Rev.* D **70** 124019
[9] M Cadoni, S Mignemi and M Serra 2011, *Phys. Rev.* D **84** 084046
[10] K Goldstein, S Kachru, S Prakash and S P Trivedi 2010, *JHEP* **1008** 078
[11] E Perlmutter 2011, *JHEP* **1102** 013
[12] S Kachru, X Liu and M Mulligan 2008, *Phys. Rev.* D **78** 106005
[13] T Hertog 2006, *Phys. Rev.* D **74** 084008
[14] P Breitenlohner and D Z Freedmann 1982, *Phys. Lett.* B **115** 197

Young Researcher Meeting in Rome 2012 IOP Publishing
Journal of Physics: Conference Series **383** (2012) 012004 doi:10.1088/1742-6596/383/1/012004

Kaonic atoms measurements at the DAΦNE collider: the SIDDHARTA experiment

A. Rizzo[1], M. Bazzi[1], G. Beer[2], C. Berucci[1], L. Bombelli[3], A. M. Bragadireanu[1,4], M. Cargnelli[5], C. Curceanu (Petrascu)[1], A. d'Uffizi[1], C. Fiorini[3], T. Frizzi[3], F. Ghio[6], C. Guaraldo[1], R. S. Hayano[7], M. Iliescu[1,4], T. Ishiwatari[5], M. Iwasaki[8], P. Kienle[5,9], P. Levi Sandri[1], A. Longoni[3], J. Marton[5], S. Okada[1], D. Pietreanu[1,4], T. Ponta[4], A. Romero Vidal[10], E. Sbardella[1], A. Scordo[1], H. Shi[7], D. L. Sirghi[1,4], F. Sirghi[1,4], H. Tatsuno[1], A. Tudorache[4], V. Tudorache[4], O. Vazquez Doce[9], E. Widmann[5], B. Wünschek[5], J. Zmeskal[5]

[1] Laboratori Nazionali di Frascati - INFN, Via Enrico Fermi 40, Frascati 00044 (Rome), Italy
[2] Department of Physics and Astronomy, University of Victoria - Victoria B.C., Canada
[3] Politecnico di Milano, Sezione di Elettronica - Milano, Italy
[4] IFIN-HH - Magurele, Bucharest, Romania
[5] Stefan-Meyer-Institut für subatomare Physik - Vienna, Austria
[6] INFN, Sezione di Roma 1 and Istituto Superiore di Sanitá - Rome, Italy
[7] University of Tokyo - Tokyo, Japan
[8] RIKEN, The Institute of Physics and Chemistry Research - Saitama, Japan
[9] Excellence Cluster Universe, Tech. Univ. Munchen, Garching, Germany
[10] University of Santiago de Compostela - Santiago de Compostela, Spain

E-mail: `alessandro.rizzo@lnf.infn.it`

Abstract. This work presents the SIDDHARTA experimental results on the kaonic hydrogen and helium atoms, reviewing some fundamental aspects of the physics case. After an introduction about the fundamental role played by these bound systems and their X-ray transitions measurements in testing the reliability of the Chiral Perturbation Theory as a realisation of Quantum Chromodynamics at low energies, a short review of the light kaonic atoms physics is given. The review, highlighting the experimental challenges, aims to better understand the SIDDHARTA performances to obtain the most precise spectroscopic X-ray measurements of these light kaonic atoms. A description of the SIDDHARTA apparatus is followed by the presentation of the results on kaonic hydrogen and helium, nowadays the most precise available in literature.

1. Introduction

The SIDDHARTA experiment at the DAΦNE collider has the aims to investigate the strong interaction between K^- and nucleons at very low energies, via the X-ray spectroscopic measurements of Light Kaonic Atoms (LKA), bound systems made of the negatively charged kaon and a light nucleus. The presence of the short range strong interaction in such systems, where the binding energies (keV range) are far below the hadronic scale (about 1 GeV), makes them the ideal probe to check the nature of the Chiral Symmetry Breaking in the strangeness sector through a real threshold measurement.

2. Kaonic hydrogen and helium: strong interaction investigation at very low energies in the strangeness sector

To account for the strong interaction in the kaonic hydrogen and helium systems, it is helpful to introduce the main processes characterising them. Following ref.[1], it is possible to recognise four main processes, namely the capture, the formation , the cascade and the nuclear absorption. After the capture of the K^- meson in an external orbital of the target atom and the formation of the new bound system, the meson starts a series of electromagnetic transitions towards the low lying levels (cascade process). After the atomic shell depletion via the Auger emission, such transitions occur mostly radiatively and in the last ones, when the K^- is close to the nucleus experiencing the strong interaction too, the photon emission is in the X-ray band. A high precision spectroscopic measurement of the lines related to these last transitions allows to evaluate the energy shift (ϵ) and the width (Γ) with respect to the pure electromagnetic case. Recognising these parameters as the real and the imaginary part of the complex K^-p S-wave scattering length a_{K^-p} using improved Deser-type formula [2][3][4][5], it is possible to relate them to different theories which describe the strong interaction at low energies. The theoretical description of the low-energy strong interaction has reached different levels, reflected in the different approaches applied to the LKA. Such are the optical potential models, used in single-particle approach to naturally insert the nuclear absorption of the meson (a sink), and the effective theories based on the chiral SU(3) symmetry [6][7]. The SIDDHARTA experiment is then ideal to test a large class of low-energy strong interaction theories, thanks to its unprecedented spectroscopic measurements precision.

3. The experimental challenging of the light kaonic atoms physics

Starting from the main processes which take place in LKA, it is possible to highlight the experimental requirements to perform a precision X-ray spectroscopy measurement of these systems. In particular, the capture and the cascade processes are those which mostly constrain the experimental method, as shown below.

The light kaonic atoms production and the capture process: One of the main experimental issues in the LKA physics is their production, a task strictly related to the capture process [8]. A very simple calculation for the kaonic hydrogen, using the thumb rule that the capture process cross section is maximum when the K^- kinetic energy is comparable with the one of the electron in the atomic capturing orbital, gives the value 13.2 keV, showing that are needed low energy kaons and a "thermalisation" process to slow them down. Only two experimental techniques are nowadays available to fulfil these requirements: an extracted beam or the use of a Φ factory, an e^+e^- collider tuned on the Φ resonance which decays in K^+K^- pair with a branching ratio of 49.2%. The main differences between these experimental techniques concern two fundamental parameters: the energy of the produced kaon beam and the signal over background ratio in the acquired spectrum. A kaon beam produced with the extracted beam technique is much more energetic than the one produced at a Φ factory, so in order to thermalise it more degrader material is required, enhancing the spread of the beam momentum distribution and the beam loss due to the decay. Moreover, the contamination by μ and π of the extracted beam contributes to have worst signal over background ratio. A Φ factory, like the $DA\Phi NE$ collider of the Laboratori Nazionali di Frascati of INFN, offers a low-background working conditions which enhance the signal over background ratio in the acquired spectrum, balancing the lower intensity of the produced beam. It allows as well, thanks to the lower kinetic energy of the produced kaons, to use thin plastic degrader for the K^- thermalisation, avoiding the beam decay and the spread of the meson momentum distribution, keeping the

monochromaticity high.
The atomic cascade and the competitor processes: The cascade starts after the formation process, where the overlap of the wave-functions of the captured K^- and of the outermost electron is maximal [1]. After the depletion of all low lying atomic orbital till a formation of a hydrogenoid system $K - nucleus$, the X-ray transitions of experimental interest take place in the last part of the cascade, when the meson is nearest to the nucleus experiencing the strong interaction too. Aiming to enhance the signal over background ratio in the acquired spectrum, it is necessary to keep as higher as possible the X-ray yields controlling the contribution of the competitor processes. These processes, generated by the interaction of the LKA with the other atoms or molecules of the target, can generate loss in the signal. They can be collisional, as the Coulomb De-excitation and Elastic Scattering, or pervasive as the Stark Mixing [9].

The Stark Mixing process is very important in the hydrogen case: the small dimensions of a kaonic hydrogen atom, a Bohr radius about a factor 10^3 smaller than the one of a standard hydrogen atom, and a zero total charge, allow to the kaonic hydrogen atom to easily pass through the orbitals of the other atoms/molecules of the target, experiencing a very intense electric field. The perturbation in the Hamiltonian due to the electric field, mixes the states of the K^-p system with $\Delta n = 0$ and $\Delta \ell = \pm 1$. When the mixing occurs, the kaon gets in a high spherical state ($n > 1, \ell = 0$) with an enhanced probability: in this situation the selection rules suppress all other transition favouring the direct nuclear absorption. In the collisional processes the energy release of a transition of the K^- goes into the kinetic energy of the two colliding objects (the LKA and one atom/molecule of the target), suppressing the radiative emission. To keep under control the contribution of these processes, considering their origin, one possible solution is to enhance the mean free path of the LKA, or in other words, to reduce the target density.
Experimentally, however, is required a fine tuning of the target density because a low target density suppresses the competitor processes lowering at the same time the LKA production and so the signal, whereas a high target density enhances the LKA production rate but increases the contribution of the competitor processes, suppressing the signal.

In the next section we present how the SIDDHARTA apparatus fulfil these requirements in order to obtain the most precise X-ray measurements of kaonic hydrogen and helium.

4. The SIDDHARTA apparatus

The ideal low-energy monochromatic K^- beam to produce kaonic atoms in the target cell of the SIDDHARTA apparatus is provided by the e^+e^- collider DAΦNE of the Laboratori Nazionali di Frascati of INFN, tuned on the Φ resonance at rest (1020 MeV). With a branching ratio of 49.2%, the Φ decays in K^+K^- emitted back to back, with a momentum of 127 MeV/c and a spread $\Delta p/p$ of 0.1%. The low kaons momentum allows to use a thin plastic degrader for the thermalisation, keeping the beam monochromaticity, whereas the particular topology of the Φ decay allows to implement a kaon trigger system to reduce the experimental background, as explained below. These characteristics, combined with a high machine luminosity of $4.5 \times 10^{32} cm^{-2} s^{-1}$ achieved through the Crab-Waist collisions technique [11], account DAΦNE collider as the ideal machine to perform this type of physics.

The SIDDHARTA trigger system performs Time Of Flight (TOF) measurements to disentangle the kaons emitted back-to-back and the Minimum Ionizing Particles (MIPs) resulting from particles lost by the circulating beams. Made of two scintillator placed immediately below and above the interaction point, the trigger systems allows a TOF separation between kaons and MIPs of about 1 ns.

In order to correlate the passage of a kaon in the kaon trigger scintillators with an X-ray emission coming from the produced LKA, to reduce the background in the spectroscopic measurement, a very fast detector for high precision spectroscopy is needed. The SIDDHARTA Silicon Drift Detectors (SDDs) combine a fast response with a good energy resolution

(FWHM=150eV@5.9KeV): 144 SDDs with 1 cm^2 of active area [12] surround the SIDDHARTA target, to detect the LKA X-rays. In the SIDDHARTA SDDs the minimisation of the anode capacitance and a JFET preamplifier directly integrated on the anode reduce the noise contribution whereas a collection-charge field in the depletion volume allows to keep a large active area. The stability of each detector has been continuously monitored during the data taking with *in-situ* calibrations, one every 4 hours, using a sandwich of Ti and Cu foils and an X-ray tube. The calibration data have been used to characterise every single detector, allowing to keep under control in the final analysis some systematics dependent on the detectors, the Front End Eletronics (FEE) and the Data AcQuisition system (DAQ). The produced K^-, passing through the scintillator of the trigger system and the plastic degrader, enters in the SIDDHARTA target cell, placed in the vacuum chamber and surrounded by the SDDs. To enhance the mean free path of the kaonic atoms in the target cell minimising the contribution of the competitor processes, a gaseus target is used. The temperature and the pressure of the target are carefully set, because the capture probability is proportional to the density of the target itself. In particular, for the different SIDDHARTA targets (H, ^{4}He and ^{3}He), the temperature and the pressure were set as follows: 23 K and 1 bar for the H [13], 20K and 1 bar for the ^{3}He [14] and 27K and 0.95 bar for the ^{4}He [15]. For a detailed description of the SIDDHARTA apparatus see ref.[16].

5. SIDDHARTA experimental results

In this section the SIDDHARTA results on kaonic hydrogen and helium are briefly reported; for a detailed description of the analysis, see ref.[13] for the KH, ref.[14] for K^3He and ref.[15] for K^4He:

- **KH**: A data set for a total amount of 340 pb^{-1} integrated luminosity has been analysed. The results for the shift ϵ and the width Γ for the K_α transition are: $\epsilon_{1s} = -283 \pm 36(stat) \pm 6(syst)\ eV$ and $\Gamma_{1s} = 541 \pm 89(stat) \pm 22(syst)\ eV$. It is the most precise result available in literature.
- **K^{4}He**: A data set for a total amount of 20 pb^{-1} integrated luminosity has been collected and analysed. A source of ^{55}Fe has been used during the data taking to check the systematic errors. The results for the shift ϵ for the L_α transition is: $\epsilon = 0 \pm 6(stat) \pm 2(syst)\ eV$. This is the first result obtained using a gaseous ^{4}He target, that confirm the result of a previous experiment with a liquid target (see ref.[17]).
- **K^{3}He**: A data set for a total amount of 15 pb^{-1} integrated luminosity has been collected and analysed. The results for the shift ϵ for the L_α transition is: $\epsilon = -1.7 \pm 2.7(stat) \pm 4(syst)\ eV$. This is the first measurement in the world for this light kaonic atom.

6. Conclusions and outlook

Thanks to the monochromatic low-energy kaon beam delivered by the DAΦNE collider, to the use of a gaseous target and of fast, high-precision, detectors for X-ray spectroscopy (SDDs), the SIDDHARTA experiment performed unprecedented measurements on kaonic hydrogen, helium 3 and 4.

Presently, an upgrade of the SIDDHARTA apparatus is ongoing for the SIDDHARTA(-2) phase. The main aim is to measure the kaonic deuterium transitions to the fundamental level, allowing to access the isospin-dependent antikaon-nucleon scattering lengths for the first time. Other measurement of LKA are considered as well, in particular the X-ray transitions to the fundamental level of the K^4He and K^3He.

The SIDDHARTA(-2) apparatus together with $DA\Phi NE$ represent indeed a unique opportunity to understand the low-energy QCD in strangeness sector in systems with few nucleons.

Acknowledgments

We thank C. Capoccia, G. Corradi, B. Dulach, and D. Tagnani from LNF-INFN; H. Schneider, L. Stohwasser, and D. Stückler from Stefan-Meyer-Institut, for their fundamental contribution in designing and building the SIDDHARTA setup. We thank as well the $DA\Phi NE$ staff for the excellent working conditions

References

[1] Gotta D 2004 *Progress in Particle and Nuclear Physics* **52** 133
[2] Meißner U-G *et al* 2004 *Eur. Phys. J.* C **35** 349
[3] Döring M, Meißner U-G 2011 *Phys. Lett.* B **704**, 663
[4] Borasoy B *et al.* 2005 *Phys. Rev. Lett.* **94** 213401
[5] Deser S *et al.* 1954 *Phys. Rev.* **96**, 774, and Truemann T L 1961 *Nucl. Phys.* **26** 57
[6] Ikeda Y*et al.* 2011 Phys. Lett. B **706** 63
[7] Borasoy B *et al.* 2005 *Eur. Phys. J.* A **25** 79
[8] Cohen J S 2004 *Rep. Prog. Phys.* **67** 1769
[9] Jensen T S, Markushin V E 2002 *Eur. Phys. J.* D **21** 261
[10] Day T B *et al.* 1959 *Phys. Rev. Lett.* **3** 61
[11] Boscolo M *et al.* 2010 *Nucl. Instr. and Meth.* A **621** 157
[12] SIDDHARTA Collaboration, Bazzi M *et al.* 2011 *Nucl. Instr. and Meth.* A **628** 264
[13] SIDDHARTA Collaboration, Bazzi M *et al.* 2011 *Phys. Lett.* B **704** 113
[14] SIDDHARTA Collaboration, Bazzi M *et al.* 2011 *Phys. Lett.* B **697** 199
[15] SIDDHARTA Collaboration, Bazzi M *et al.* 2009 *Phys. Lett.* B **681** 310
[16] SIDDHARTA Collaboration, Bazzi M *et al.* 2012 *Nucl. Phys.* A **881** 88
[17] Hayano R S 2011 *Journal of Physics: Conference Series* **312** 022004

Journal of Physics: Conference Series **383** (2012) 012005 doi:10.1088/1742-6596/383/1/012005

Measurement of the Z/γ^* forward-backward asymmetry in muon pairs with the ATLAS experiment at the LHC

Giulio Cornelio Grossi

Universitá degli studi di Roma "Tor Vergata" and INFN Roma2

E-mail: giulio.grossi@roma2.infn.it

Abstract. A study on muon pairs produced through an intermediate Z/γ^*, in pp collisions at the LHC, at a center-of-mass energy of 7 TeV, is presented. After a selection aimed at enhancing the contribution from Z boson decay, the topology of the events is analyzed, and the forward-backward asymmetry is measured. The result is then used to test a method to measure the effective weak mixing angle $\sin^2\theta_W^{eff}$. This note summarizes the results obtained with 2011 data collected by the ATLAS experiment at the LHC, corresponding to an integrated luminosity of 4.8 fb^{-1}.

1. Introduction

The electroweak theory represents a fundamental component of what is believed to be the theory that describes all the particles and their interactions, the Standard Model (SM). Despite the great success of the Standard Model, it is not believed to be the final theory that describes particle interactions. Many open questions remain, i.e. the verification of mass generation by spontaneous symmetry breaking and the origin of the so called dark matter, which is a major component of the universe and cannot be explained in terms of known particles.

The Large Hadron Collider (LHC) at CERN, as the largest proton-proton collider in the history of particle physics, is designed to answer these questions. Since its first physics run on 20 November 2009 at 900 KeV, it has reached a center of mass energy of 7 TeV in 2010 and by November 2011 an integrated luminosity of about 5 fb^{-1} has been recorded by the ATLAS experiment. The integrated luminosity recorded at LHC has already allowed to perform several tests of the Standard Model.

In ATLAS, the process $pp \rightarrow l^+l^- + X$, with two leptons in the final state, is currently produced with an high rate. This process is mediated primarily by virtual photons at low values of di-lepton invariant mass (m_{ll}) [1], primarily by the Z at $m_{ll} \sim m_Z$, and by a combination of photons and Z bosons outside these regions. The presence of both vector and axial-vector couplings of electroweak bosons to fermions in the process $q\bar{q} \rightarrow Z/\gamma^* \rightarrow l^+l^-$ gives rise to a forward-backward asymmetry (A_{fb}) in the polar angle of the lepton momentum relative to the incoming quark momentum in the rest frame of the dilepton pair.

In this note a measurement of the forward-backward asymmetry with the ATLAS experiment, using $pp \rightarrow Z/\gamma^* \rightarrow \mu^+\mu^-$ events at $\sqrt{s} = 7$ TeV, is presented. The analysis focuses on the decay of the Z boson into two muons, since this provides a relative clean signature which can be clearly discriminated from other processes. Section 2 focuses on the definition of the forward-backward asymmetry and the Collins-Soper reference frame. In Section 3 I present the analysis to select the Z/γ^* candidates and the measurement of A_{fb} as a function of the di-muon invariant mass. In Section 4, a method based on a fit on the asymmetry distribution is proposed, in order to extract the weak mixing angle $\sin^2\theta_W^{eff}$. A Monte Carlo closure test is preformed, and a quite good agreement with the Monte Carlo default value si observed.

2. The Forward-Backward charge asymmetry

At LHC, the Drell-Yan process is $q\bar{q} \to Z/\gamma^* \to \mu^+\mu^-$; the lowest order proceeds via photon and Z boson exchange. The neutral current couplings of a fermion f to the Z boson has vector and axial-vector components: $J_{Zf} = \bar{f}(g_V^f + g_A^f\gamma_5)f$, where g_V^f and g_A^f are vector and axial-vector couplings of the fermion to the Z respectively.

The differential cross section for fermion pair production around the Z-pole can be written as:

$$\frac{d\sigma(q\bar{q} \to \mu^+\mu^-)}{d\cos\theta} = C\frac{\pi\alpha^2}{2s}[Q_\mu^2 Q_q^2(1+\cos^2\theta) + Q_\mu Q_q Re(\chi(s))(2g_V^q g_A^\mu(1+\cos^2\theta) + +4g_A^q g_A^\mu) + |\chi(s)|^2((g_V^{q\,2} + g_A^{q\,2})(g_V^{\mu\,2} + g_A^{\mu\,2})(1+\cos^2\theta) + 8g_V^q g_A^q g_V^\mu g_A^\mu \cos\theta)] \tag{1}$$

where C is the color factor, θ is the emission angle of the muon (anti-muon) relative to the quark(anti-quark) in the rest frame of the muon pair, and

$$\chi(s) = \frac{1}{\cos^2\theta_W \sin^2\theta_W}\frac{s}{s - M_Z^2 + i\Gamma_Z M_Z} \tag{2}$$

The first and the third terms in Equation 2 correspond to the pure γ^* and Z exchange respectively while the second term corresponds to the Z/γ^* interference. The angular dependence of the various term is either $\cos\theta$ or $(1+\cos^2\theta)$. The $\cos\theta$ terms integrate to zero in the total cross section but induce the forward-backward asymmetry.

The differential cross section in Equation 2 simplified into:

$$\frac{d\sigma}{d\cos\theta} = A(1+\cos^2\theta) + B\cos\theta \tag{3}$$

where A and B are functions that take into account the weak isospin and charge of the incoming quarks and the transfered momentum Q^2 of the interaction. Events with $\cos\theta > 0$ are called forward events, and events with $\cos\theta < 0$ are called backward events. The integrated cross section for forward events is thus $\sigma_F = \int_0^1 \frac{d\sigma}{d\cos\theta}d\cos\theta$ and the integrated cross section for backward events is $\sigma_B = \int_{-1}^0 \frac{d\sigma}{d\cos\theta}d\cos\theta$. The forward-backward charge asymmetry A_{FB} is defined as

$$A_{FB} = \frac{\sigma_F - \sigma_B}{\sigma_F + \sigma_B} = \frac{\int_0^1 \frac{d\sigma}{d\cos\theta}d\cos\theta - \int_{-1}^0 \frac{d\sigma}{d\cos\theta}d\cos\theta}{\int_0^1 \frac{d\sigma}{d\cos\theta}d\cos\theta + \int_{-1}^0 \frac{d\sigma}{d\cos\theta}d\cos\theta} = \frac{N_F - N_B}{N_F + N_B} = \frac{3B}{8A} \tag{4}$$

where N_F and N_B are numbers of forward and backward events.

When the incoming quarks participating in the Drell-Yan process have no transverse momentum relative to their parent baryons, θ is determined unambiguously from the four-momenta of the muons by calculating the angle that the muon makes with the proton beam in the center-of-mass frame of the muon pair. When either of the incoming quarks has significant transverse momentum, however, there exists an ambiguity in the four-momenta of the incoming quarks in the frame of the di-muon pair, since one can not determine the four-momenta of the quark and antiquark individually. The Collins-Soper formalism[3] is adopted to minimize the effects of the transverse momentum of the incoming quarks. In this formalism, the polar axis is defined as the bisector of the proton beam momentum and the negative of the anti-proton beam momentum when they are boosted into the center-of-mass frame of the di-muon pair, which is shown in Figure 1. The variable θ^* is defined as the angle between the muon and the polar axis. Let $Q(Q_T)$ be the four momentum (transverse momentum) of the dimuon pair, P_1 and P_2

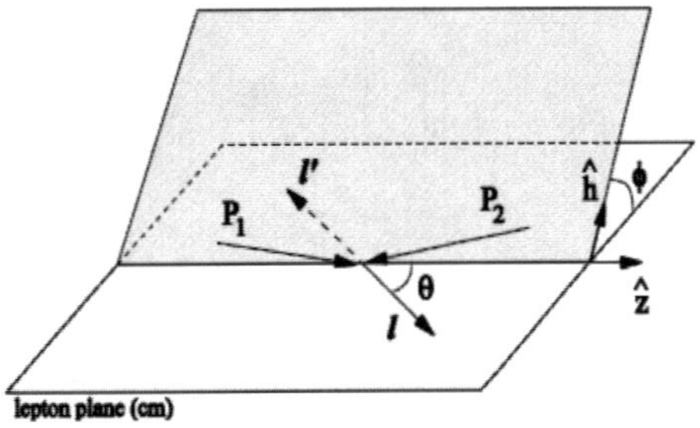

Figure 1. The Collins-Soper reference frame

be the four-momentum of the muon and anti-muon respectively, all measured in the lab frame. Then $\cos\theta^*$ is given by

$$\cos\theta^* = \frac{2}{Q\sqrt{Q^2 + Q_T^2}}(P_1^+ P_2^- - P_1^- P_2^+) \tag{5}$$

where $P_i^\pm = \frac{1}{\sqrt{2}}(P_i^0 \pm P_i^3)$, and P^0 and P^3 represent energy and longitudinal component of the momentum.

3. Measurement and unfolding of the A_{fb}

The event topology of $pp \to Z/\gamma^* \to \mu^+\mu^+$ has a very characteristic signature: two high energetic and isolated muons in the final state are produced. The data sample used in this analysis was collected using the ATLAS detector between March 21 and October 30. The number of events in the data correspond to an integrated luminosity of 4.8 fb^{-1}. The way to select only interesting events is to use the characteristic signature of the signal process. Some constraints on the kinematic quantities of muons in data sample are applied and, at the end, an almost pure $Z/\gamma^* \to \mu^+\mu^+$ sample is obtained. After the application of the selection requirements we found 1.3M Z/γ^* candidates in data sample. The final sample is still contaminated by some background events surviving the selection but this contamination is expected to be very low because, as already mentioned, the signal process is characterized by a very clear signature. The same selection cuts are also applied to Monte Carlo signal and to the background samples.

The selection of $Z/\gamma*$ candidates allows to measure some direct observables related to the process $Z/\gamma^* \to \mu^+\mu^+$. The forward-backward asymmetry A_{fb} of the muons pair is one of the observables that can be used to test the validity of the electroweak theory and, at the same time, to search for new physics signatures. A measurement of the forward-backward asymmetry can be performed following the steps below:

- Divide the mass spectrum in a certain number of bins.
- In each bin count the number of forward and backward events.
- Subtract the number of forward and backward events due to the background.

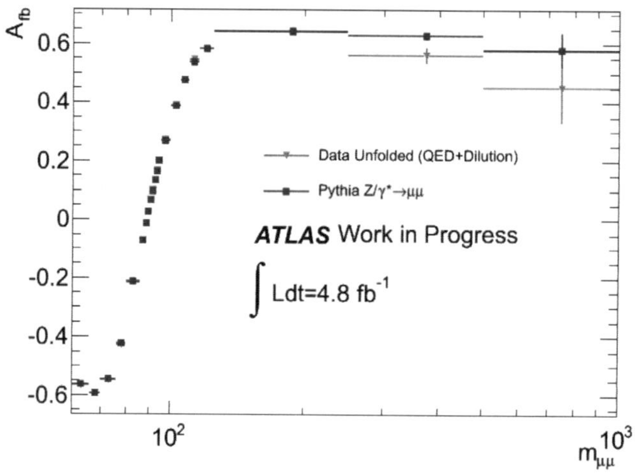

Figure 2. Spectrum of measured A_{fb}. Squares refer to raw A_{fb} spectrum. Circles refer to unfolded A_{fb} spectrum.

- Compute the A_{fb} value using the relation:

$$A_{fb} = \frac{N^F - N^B}{N^F + N^B} \tag{6}$$

We divide an invariant-mass range, from 60 to 1000 GeV, in 21 bins. We apply the measurement procedure to the Z/γ^* candidates found in data sample and obtain a distribution of *raw* A_{fb} vs. $m_{\mu\mu}$.

The measured spectrum of the asymmetry needs to be corrected for three main effects: radiative corrections, detector resolution and dilution. The effect of radiative corrections and detector resolution [4] is mainly to deform the shape of the Z/γ^* mass spectrum; this effect is called *mass migration*. Photon radiation of the muons lowers the di-muon invariant mass. Events from the Z peak region are therefore shifted towards smaller values of $m_{\mu\mu}$, thus reducing the cross section in and above the peak region, and increasing the rate below the Z pole. Since events from the Z peak, where A_{fb} is positive and small, are shifted towards smaller values of $m_{\mu\mu}$ by photon radiation, the forward-backward asymmetry is significantly reduced in magnitude by radiative corrections. The other important correction is called *dilution*, an effect directly related to the direction of the incoming quark. In *pp* collisions, there is in principle no way of knowing which one of the beams contributed a quark to the collision. This effect gives rise to a sensible reduction of the A_{fb}.

The measurement of A_{fb} is corrected for this effect by means of a response-matrix based unfolding. Using the available Monte Carlo $Z/\gamma^* \rightarrow \mu\mu$ samples, we can calculate the probability that an event with a given *true* $m_{\mu\mu}$ will be reconstructed with a different value of $m_{\mu\mu}$. This matrix is then applied to the observed spectrum (and A_{fb}) to the one at Born level.

Also dilution correction is based on response matrices. The unfolding procedure is the same as the mass-migration correction. By the fact that dilution deforms the $\cos\theta^*$ distribution, we can consider the *true* $\cos\theta^*$ of a given event and calculate the probability that *true* $\cos\theta^*$ will be reconstructed with a different value. The resulting spectrum of A_{fb} is showed in figure 2. No particular deviation from the SM prediction is observed.

4. Extraction of $\sin^2\theta_W^{eff}$

The measurement of the asymmetry presented in the Section 3 can be interpreted as a measurement of $\sin^2\theta_W^{eff}$. In fact the axial and axial-vector couplings that appear in the expression of A_{fb} are directly related with the value of $\sin^2\theta_W^{eff}$ by the relations

$$g_V^f = I_3^f - 2Q_f \sin^2\theta_{eff}^f \quad (7)$$

$$g_A^f = I_3^f \quad (8)$$

In order to extract a measurement of $\sin^2\theta_{eff}^f$ from the unfolded A_{fb} spectrum we use an expansion of A_{fb} in terms of the center of mass energy, around the Z pole[5]

$$A_{fb}(s) \simeq A_{fb}\left(m_Z^2\right) + \frac{(s-m_Z^2)}{s}\frac{3\pi\alpha(s)}{\sqrt{2}G_F m_Z^2}\frac{2Q_q Q_f g_{Aq} g_{A\mu}}{\left(g_{Vq}^2+g_{Aq}^2\right)\left(g_{V\mu}^2+g_{A\mu}^2\right)} \quad (9)$$

This expansion can be used to determine the value of $\sin^2\theta_W^{eff}$ by fitting the A_{fb} vs. $m_{\mu\mu}$ distribution in the vicinity of the Z pole. In order to test the validity of the fitting procedure, a closure test on the *true* Monte Carlo sample has been performed. The Monte Carlo default value of the weak mixing angle is $\sin^2\theta_W^{eff} = 0.232$. The result of the fitting procedure applied to the *true* A_{fb} distribution should be in agreement with this default value of the weak mixing angle. The value of $\sin^2\theta_W^{eff}$ extracted from the *true* A_{fb} distribution around the Z pole is

$$\sin^2\theta_W^{eff} = 0.23202 \pm 0.00043 \quad (10)$$

5. Conclusions

The study of $pp \to Z/\gamma^* \to \mu^+\mu^-$ process played an important role in this two years of data taking at LHC. The clear signature, given by the two isolated muons in the final state, allows to easily discriminate the signal from the background and the cross section of about 1 nb provides a large yield of Z bosons relative to the luminosity reached by LHC.

In this work, a sample of about 4.8 fb^{-1} was analyzed. After the application of kinematic cuts, about 1.3M Z/γ^* candidates were found. The candidates have been used to obtain a measurement of A_{fb}. The A_{fb} spectrum has been corrected for the effects of radiative corrections and dilution. No particular deviation from the SM prediction is seen. A method to extract the weak mixing angle $\sin^2\theta_W^{eff}$, based on a fit on the unfolded A_{fb} distribution, has been tested with a Monte Carlo closure test. The exctrated value of $\sin^2\theta_W^{eff}$ is in agreement with the default value of the Monte Carlo sample.

6. References

[1] S. D. Drell and T.-M. Yan. Phys. Rev. Lett. 25, 316 (1970).
[2] K. Nakamura *et al* (Particle Data Group). J. Phys. G: Nucl. Part. Phys. **37**, 075021 (2010).
[3] J.C. Collins and D.E. Soper. Phys. Rev. D **16**, 2219 (1977).
[4] U. Baur and S. Keller and W.K. Sakumoto. QED radiative corrections to Z boson production and the forward-backward asymmetry at hadron colliders. Phys. Rev. D **57**, (1998).
[5] R. Tenchini C.Verzegnassi, The Physics of the Z and W Bosons. USA: World Scientific (2008) p73.

Young Researcher Meeting in Rome 2012 IOP Publishing
Journal of Physics: Conference Series **383** (2012) 012006 doi:10.1088/1742-6596/383/1/012006

The ALTEA experiment onboard the International Space Station

L. Di Fino, F. Agostini, M. Larosa, L. Narici and V. Zaconte

Department of Physics, University of Rome "Tor Vergata", Rome (Italy)

E-mail: `luca.difino@roma2.infn.it`

Abstract. The knowledge of the composition of the radiation environment is an important information for all the radiation safety issues needed for the planning of future long manned space missions. The ALTEA detector is on board the International Space Station since July 2006 and during this period it has performed a detailed measurement of the radiation environment. In this paper we present a summary of past measures and results.

1. Introduction

The detailed knowledge of the radiation environment in the International Space Station (ISS) is a mandatory step toward a successful radiation risk assessment, needed for the planning of future long manned space missions. ALTEA (Anomalous Long Term Effects on Astronauts) is a multidisciplinary program also devoted at monitoring the International Space Station radiation environment. The ALTEA active particle telescopes feature real time monitoring capability and nuclear discrimination [1, 2, 3]. The ALTEA-space detector was placed on board the ISS on July 2006 and it has been operative since August 2006 in different configurations and different positions and orientations inside the UsLab and the Columbus module of the International Space Station (ISS). In this paper we will provide a summary of the results and findings of these years of measurements.

2. The ALTEA-space detector

The ALTEA cosmic ray detector is composed by six Silicon Detector Units (SDUs). Each SDU consists of a particle telescope made of six planes divided in strips to measure energy loss and direction of detected particles. Each SDU has a geometric factor of 230 $cm^2\, sr$. The LET (in silicon) range of the detector goes from a threshold of $2.9 \pm 0.2\ keV/mm$ up to about 800 keV/mm, which corresponds to ions between Li and Fe with a small energy windows on Helium and protons. The modularity of ALTEA permits to work with different configurations. The SDUs were arranged on a helmet-shaped structure until September 2010. In this configuration the whole detector sports a geometric factor of 1190 $cm^2\, sr$. For further details the reader is addressed to [1].

In this configuration ALTEA can operates in two different modes, named Dosimetry (DOSI) and Central Nervous System Monitoring (CNSM), which are, respectively, unmanned and manned operative modes. DOSI is a continuous measure, while the CNSM sessions lasts 90 minutes. During DOSI mode ALTEA measures radiation fluxes on board ISS. Only Silicon

Detector System (SDS) is switched on and telemetry data, both engineering and scientific, are downloaded in real-time. Seven $90 - min$ CNSM sessions on three different astronaut subjects were performed between 2006 and 2007. During each session the astronaut signal perceptions of Light Flashes with a pushbutton while SDS will measure cosmic rays flux, so that it is possible to study concurrently the passage of cosmic radiation through the brain, the functional status of the visual system and the electrophysiological dynamics of the cortical activity.

In September 2010 ALTEA SDUs were rearranged in a different configuration on a different support structure for Survey, the first part of the ESA experiment ALTEA-Shield. Each SDU couple forms a silicon telescope of twelve planes, and each telescope aims at a different direction in space. In this configuration ALTEA measured the cosmic ray particle flux until June 8, 2012, when ALTEA was moved to the Columbus module of the ISS to start the second part of the ALTEA-Shield experiment, named Shield.

3. Measurements
The typical particle rate measured by the ALTEA SDS shows the typical periodic modulations due to the passage between high and low latitude regions (see Fig.1)[2].
The highest peaks correspond to passes through the South Atlantic Anomaly (SAA), where there is a large increase of the particle flux due to the contribution of the protons trapped in the inner radiation belt. The lower curve shows the particle count rate of the $Z > 5$ component which, excluding protons, does not show SAA increase. There is another visible modulation

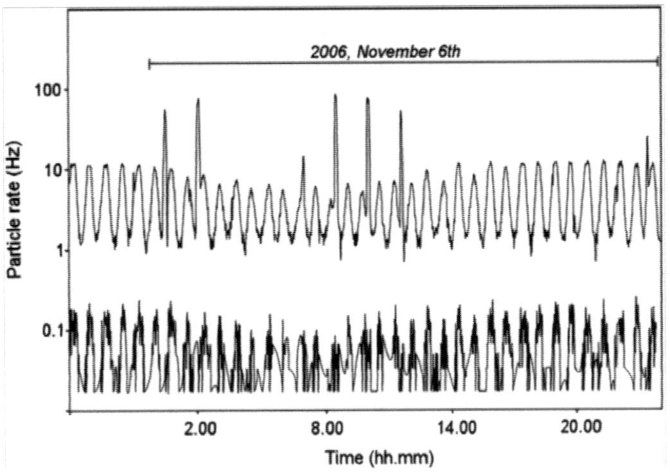

Figure 1. Particle rate for 1 day of ALTEA data, the typical oscillations due to the passage between high and low latitude regions and the SAA high peaks are visible. The $24-h$ modulation is the longitude effect [2].

within a $24 - h$ period called the longitude effect, which is due to the combination of the orbital precession and the different positions of the geographic and geomagnetic poles. Filtering the fast particles with a selection on the energy loss difference between the first and the last couple of silicon planes to be less than 10% we obtain a typical spectrum where the elements from Boron ($Z = 5$) to Iron ($Z = 26$) are clearly discernible.

Correlating ALTEA data with ISS orbital information, we grouped these events with respect to the value of their geomagnetic coordinates [3], B and L, B being the magnetic field value

Journal of Physics: Conference Series **383** (2012) 012006 doi:10.1088/1742-6596/383/1/012006

and L the McIlwain coordinate. Three regions were considered: the polar region or region of Galactic Cosmic Rays (GCR), the South Atlantic Anomaly and the equatorial region. In Fig. 2 the three time normalized spectra are plotted. Relative nuclear abundances, calculated with the method described in [2] starting from a Landau fit to reconstruct the full energy released spectra, are presented in the three different regions and compared with the abundances found for the total data (see Fig. 3). ALTEA was able to measure the particle rate and energy spectrum

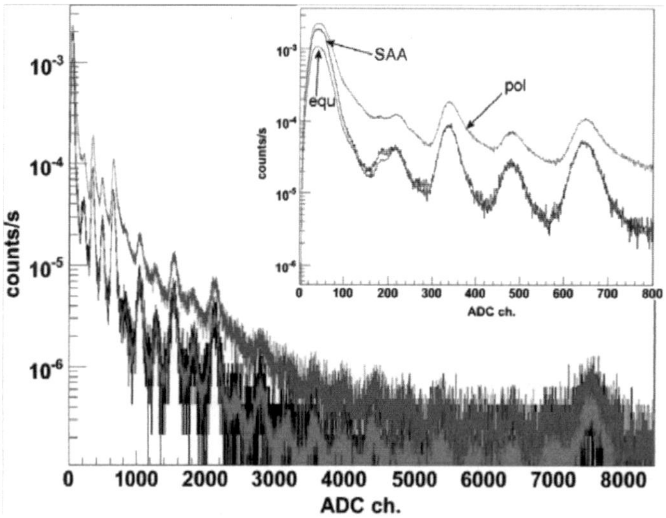

Figure 2. Energy spectra of data acquired in three different geomagnetic regions: in the inbox, the zoomed plot at the lower energies [3].

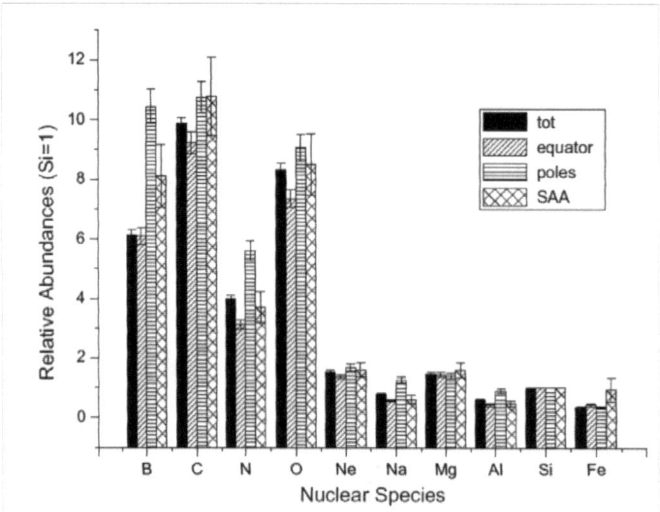

Figure 3. Relative nuclear abundances ($Si = 1$) for the different geomagnetic regions and for the total data [3].

during the December 2006 SPEs [4]. The December 2006 SPEs, occurred between 6th and 14th, originated from the active region NOAA 10930. The comparison with the ($> 50\ MeV$) proton flux measured by the GOES-11 satellite shows the correlation of the events recorded with the two instruments. The SPE effects can finally be identified in the unusually high rate peaks of Fig. 4. A study on ion abundances was performed using a novel algorithm that permits to estimate nuclear composition even from spectrum that lacks enough statistics to apply the standard fitting method [5]. Comparing these measurements with analogous measurements during quiet Sun conditions we observed an increase of the light ion ($Z < 5$) rate with respect to the values recorded during the quiescent times. In our limited acceptance window for proton the mean measured rate over the whole SPE week is 1.5 times greater than the rate during the quiescent condition, but the ratio rises to about 10 (lower limit) when averaged over the solar most active period (three ISS orbits). By contrast, the abundance of heavier nuclei results unchanged within the errors with respect to quiet Sun conditions. Measurements carried on during 2009

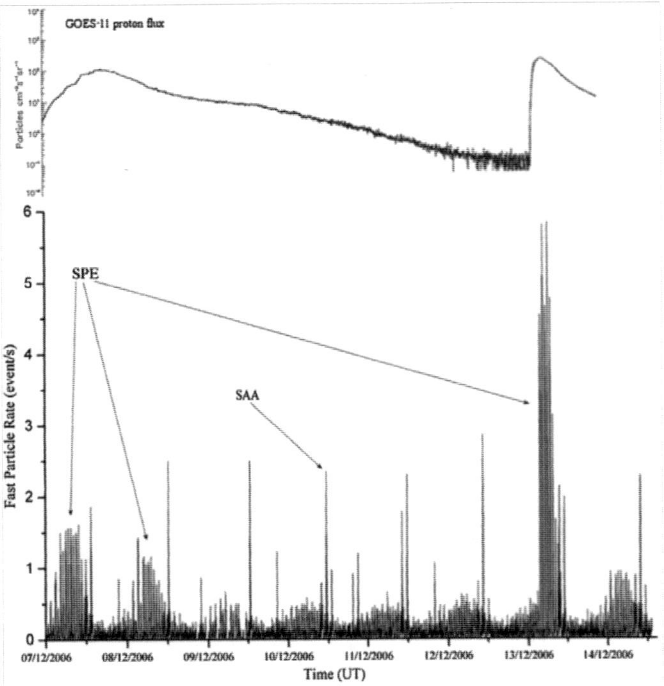

Figure 4. SDU1 rate for fast particles during the SPE week in comparison with the GOES-11 ($> 50\ MeV$) proton flux during the same period of time. Each day starts from $00:00\ am$ UT. In the figure the longitude effect, the passages of the ISS over the SAA and the SPE effects are well visible [4].

show that the radiation fluences in the USLab-ISS feature quite different constituents along the three axes: the radiation coming from different directions has significantly different quality. This anisotropy for high-LET particles (LET $> 50\ keV/mm$) is a factor of 3 in LET and consequently in dose and is minimum along the USLab longitudinal axis and maximum normal to it.

To analyze flux anisotropies, we calculated the LET spectra in three orthogonal directions corresponding to the main axis of ISS, where the Y and Z directions correspond to the transverse sections of the USLab while X is along the ISS longitudinal axis. All the spectra are then normalized with the proper geometric factor. We notice that high-LET particle flux is very attenuated along the X direction (Fig. 5), while at low LET ($< 13\ keV/mm$) the X spectrum is the highest (Fig. 5, inset). Integrating these LET spectra for all particles (LET $> 3\ keV/mm$)

and for high-LET particles, defined here as those with energy release more than 50 keV/mm in Silicon (corresponding to ions that deposit more energy than a Mg ion at 2 GeV/nucleon), we can compare the particle fluxes and total LET rates in the three directions defined above and in the different geomagnetic regions (see Fig. 6). High-LET particles in the X direction are a factor between 1.6 and 2.3 lower than in the Y and Z directions in all regions. This anisotropy for high-LET particles becomes a factor between 2.3 and 3.2 in terms of LET rates.

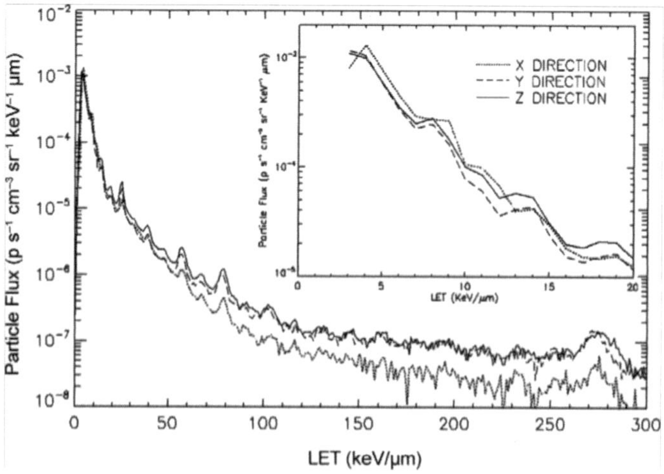

Figure 5. LET spectra in three orthogonal directions. The Y and Z directions correspond to transverse sections of ISS while X is along the ISS longitudinal axis. The X spectrum is lower than those for Y and Z for high-LET particles and higher at low LET (see inset)[6].

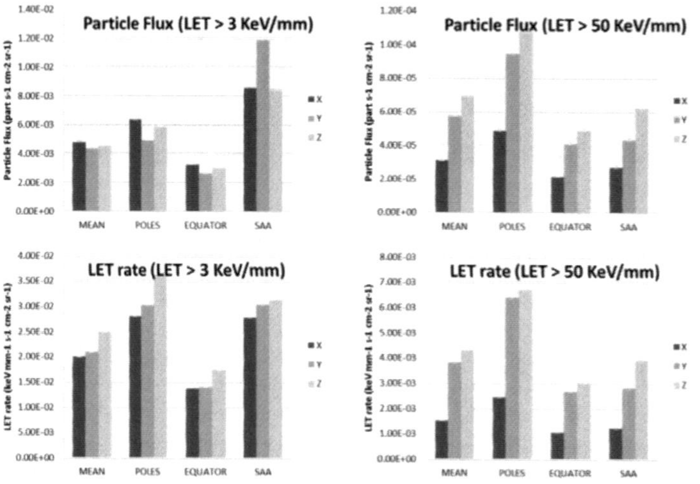

Figure 6. Particle Fluxes and LET Rates for Particle with LET Greater than 3 and 50 keV/mm. All fluxes are presented for the X, Y and Z directions and for all geomagnetic regions.

4. Conclusions

The ALTEA detector is on board the International Space Station since July 2006; it was operative non continuously within the USLab since August 2006 and it was recently moved

in the Columbus module.
In this period ALTEA, exploiting its large energy window acceptance, performed a detailed measurement of the radiation environment on board a space vessel, studying for example the different components of the cosmic rays depending on the measurement position and on the incidence direction, and it was capable to evaluate the enhancement of the flux during the SPEs.
ALTEA is now operative in the Columbus module, where it will evaluate the effects of different shielding materials on the incidence particle flux.
All the ALTEA measures will provide a hint to assess the radiation risks for the human permanence in the space.

References

[1] V. Zaconte, F. Belli, V. Bidoli, M. Casolino, L. Di Fino, L. Narici, P. Picozza, A. Rinaldi, W. G. Sannita and D. Schardt, ALTEA: The instrument calibration. *Nucl. Instrum. Methods Phys. Res. B* 266, 20702078 (2008).
[2] V. Zaconte, L. Di Fino, C. La Tessa, M. Larosa, L. Narici and P. Picozza, High energy radiation fluences in the ISS-USLab: ion discrimination and particle abundances. *Radiat. Meas.* 45, 168172 (2010).
[3] V. Zaconte, M. Casolino, C. De Santis, et al. The radiation environment in the ISS-USLab measured by ALTEA: spectra and relative nuclear abundances in the polar, equatorial and SAA regions. *Adv. Space Res.* 46, 797799 (2010).
[4] M. Larosa, F. Agostini, M. Casolino, et al. Ion rates in the international space station during the December 2006 solar particle event. *J. Phys. G: Nucl. Part. Phys.* 38, 095102 (2011).
[5] L. Di Fino, V. Zaconte, A. Ciccotelli, M. Larosa, L. Narici, Fast Probabilistic Particle Identification algorithm using silicon strip detectors, *Advances in Space Research* 50, 408414 (2012).
[6] L. Di Fino, M. Casolino, C. De Santis, et al. Heavy-Ion Anisotropy Measured by ALTEA in the International Space Station, *Rad. Res.* 176-3, 397-406 (2011).

Young Researcher Meeting in Rome 2012 IOP Publishing
Journal of Physics: Conference Series **383** (2012) 012007 doi:10.1088/1742-6596/383/1/012007

The SNR W44: an ideal laboratory for Cosmic-Ray acceleration

M. Cardillo[1,2], A. Giuliani[3] and M. Tavani[1,2,4], for the AGILE collaboration.

[1] INAF/IAPS-Roma, Via del Fosso del Cavaliere 100, 00133, Roma, Italy
[2] Department of Physics, Univ. di Roma "Tor Vergata", via della Ricerca Scientifica 1, 00133, Roma, Italy
[3] INAF-IAFS Milano, via E.Bassini 15, 20133, Milano, Italy
[4] Consorzio Interuniversitario Fisica Spaziale (CIFS), Villa Gualino, V.la Settimio Severo 63, 00133, Torino, Italy

E-mail: mcardillo@roma2.infn.it

Abstract. W44 (G34.7-0.4) is one of the most studied Supernova Remnants (SNRs) because it is an ideal system to study Cosmic-Ray (CR) production, acceleration and propagation. In the last years, SNR study showed an increasingly complex scenario with a continuous rielaboration of theoretical models; moreover, until now, providing an experimental unambiguous proof of the CR origin has been elusive, despite many decades of attempts and controversial analysis. In this context the AGILE γ-ray satellite has an important role. During its five years of life it observed a great amount of different astrophysical sources, including SNRs. In case of W44 AGILE observed a spectrum extending to energies below $E = 200$ MeV that allows us to exclude leptonic emission as the main contribution to the γ-ray emission [36]. Moreover, refined AGILE data show that W44 spectrum could be the sum of two (or more) contributions from two different regions of the remnant where different processes influence CR acceleration and propagation. Future AGILE data will lead us to understand the intricate link between SNRs and CRs.

1. Introduction

Understanding Cosmic-Ray (CR) origin is one of the most important issues of high-energy astrophysics [11, 22, 25, 31, 34, 49]. In the last few decades, scientific community agrees that CRs with energies up to 10^{15} eV are produced by galactic sources, mostly by Supernova Remnant (SNR) generated shocks. This is mainly because, on one hand, the energy throwput, since about 10% of SN explosion energy can be converted into CR kinetic energy [23, 34] and on the other hand, the presence of high magnetic fields ($\sim 100\,\mu G$) and strong X-ray variations in the SNR environments.

The main mechanism for converting kinetic SN explosion energy into CRs is the energy dependent linear Diffusve Shock Acceleration (DSA) [20, 21, 24], that provides a simple power-law spectrum of accelerated CRs $N(E) \propto E^{-\gamma}$ with an index $\gamma = 2$. This spectrum is modified by diffusive propagation with a diffusion coefficient $D(E) \propto E^{\delta}$ and $\delta = 0.7$ because CR spectrum is well fitted by a power-law with an index $\alpha = 2.7 = \gamma + \delta$.

In order to test the SNR hypothesis there are two main issues to be considered: (1) finding the low-energy spectral signatures of the hadronic contribution in the γ-ray spectrum (CRs are

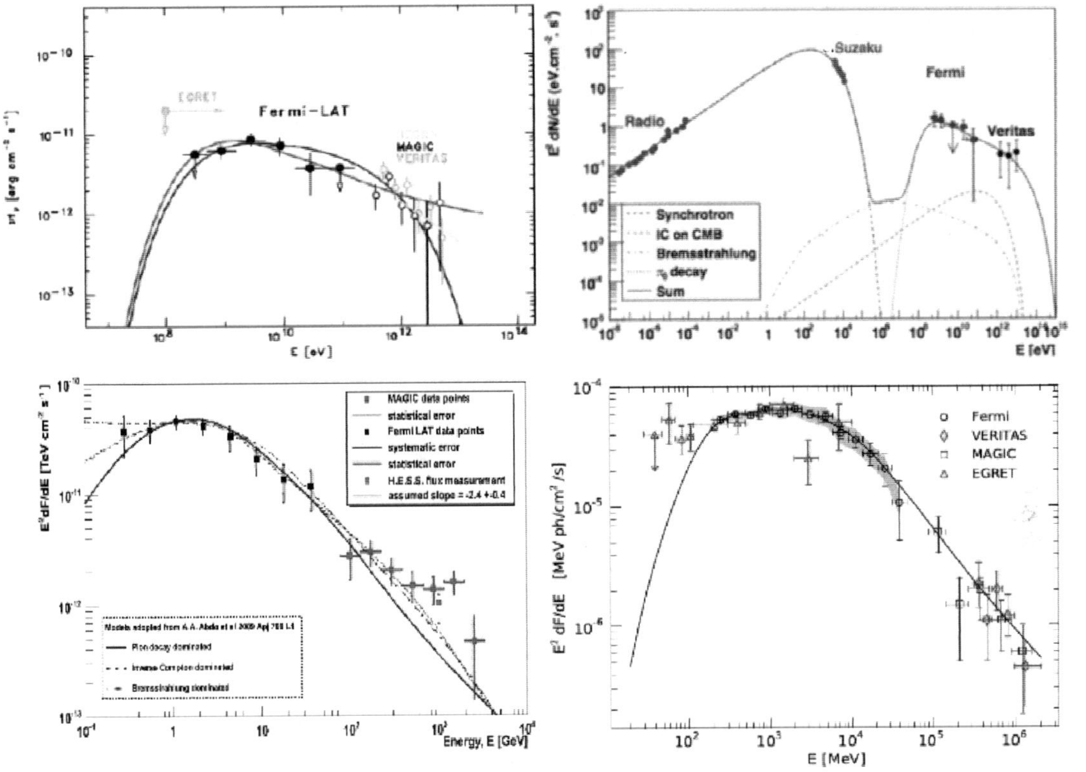

Figure 1: Spectra of four different SNRs where we highlight the lack of data at energies below 200 MeV (red solid line). **Upper left**: the youngest SNR known until now, Cas A, ~300 years old [2]; **Upper right**: Tycho SNR, about 440 years old [35]; **Bottom left**: the SNR W51c, more 20.000 years old [14];**Bottom right**: the SNR IC443, ~ 30.000 years old [4].

mainly protons),(2) observing SNR experimental spectra and understanding acceleration and propagation physical processes.

AGILE and Fermi-LAT γ-ray satellites, together with VHE instruments (HESS, VERITAS and MAGIC), provided a great amount of data from SNRs [1, 2, 4, 5, 3, 6, 7, 8, 9, 10, 12, 13, 14, 35, 37, 47] and in most cases, their spectra are well fitted by a hadronic model but we can not exclude yet the leptonic contribution (Bremsstrahlung and Inverse Compton scattering) as the main one in γ-ray emission. This is shown by experimental data below $E = 200$ MeV where hadronic and leptonic emission spectra have a well distinct behavior due to a steepening of the hadronic spectrum missing in the leptonic one (fig. 1).
Moreover all experimental SNR γ-ray emission data provide spectra steeper than those expected by a linear DSA theory especially for young SNRs (Tycho and Cas A). This feature represents a challenge even for a more detailed and realistic NLDSA theory [41] that seems to predict a concave high energy spectrum because of an highest energy hardening [15, 16].
In this context, the SNR W44 is a very interesting source. During its observations, AGILE for the first time found spectral points below $E = 200$ MeV that allowed us to exclude the leptonic emission as the main contribution to the γ-ray spectrum [36]. Moreover AGILE results confirm the steep spectral index found by Fermi-LAT; in [43] Malkov et al. try to explain this feature with the so called Alfvèn wave damping (see section 3) showing that the predicted model can well

fit W44 spectrum. However, a more detailed analysis of W44 region with AGILE seems to show a more complicated scenario with the possibility that the spectrum is due to different contributions from different SNR regions. On the basis of this consideration and of spectral steepening of the other SNRs, we are studying a theoretical context that could explain W44 peculiar spectrum.

2. W44

2.1. General Issues

W44(G34.7-0.4) is one of the most studied SNRs. It is located in the Galactic disk at a distance of about 3 kpc from Earth, and it is a middle-aged SNR, 20.000 years old [29, 52]. Radio [28] and X-ray [51] data show the mixed-morphology of the remnant, centrally peaked in the X-ray band with a well defined radio shell. On the Southern side of W44 there is the pulsar PSR B1853+01 associated with the remnant [52]. The environment of W44 is very interesting because it has an high average density [44]; this is an important requirement to enhance the proton-proton interaction probability $t_{pp} = (c\sigma_{pp}n)^{-1} \simeq 10^8\,\text{yrs} > t_{SNR} \sim 440\,\text{yrs}$ and so to detect hadronic emission from π^0 decay.

In the Northern part of the remnant there is a perfect correlation between Hα and [SII] emission [45]: [SII] traces hot-shocked gas in correspondence with the encounter between expanding shell and InterStellar Medium (ISM) density enhancement regions [33], indicating the presence of a radiative shock.

In the South-Eastern side of the remnant, instead, the presence of a molecular cloud complex and interaction between it and the SNR shell were confirmed by different analysis in the CO band [44, 45, 46, 53] and by maser OH emission [30, 38]. In the high-energy band Fermi-LAT measured a γ-ray spectrum for $E > 200$ MeV well fitted by a broken power-law hadronic distribution with a spectral index $\gamma = 3$ above the break energy E_{br} [3].

2.2. The AGILE results

The AGILE results give even more relevance to W44. The detected flux above 400 MeV is $F = (16.0 \pm 1.2) \times 10^{-8}$ ph cm^{-2} s^{-1}. Figure 2 shows that γ-ray emission detected by AGILE follows the quasi-elliptical pattern of the interior of the radio shell (green contours). There is a very good correlation with radio brightness enhancement on the North-West and South-East that are respectively in coincidence with [SII] emission and with the molecular cloud [36]. This correlation leads us to consider γ-ray emission due to CR proton acceleration by SNR shocks, confirming Fermi-LAT hypothesis.

The W44 spectrum measured by AGILE is even much more interesting (fig. 2b): comparing it with other SNR spectra in fig.1, it is evident that, for the first time, we have data at energies below 200 MeV and we can observe the low-energy steepening peculiar to hadronic γ-ray emission, being the fundamental proof of CR acceleration. A multiwavelength analysis taking into account radio data [28], TeV upper limits [3] and the previous medium density estimations, together with our data, allowed us to exclude all possible leptonic models, Bremsstrahlung or inverse Compton dominated. Assuming that synchrotron radio and γ-ray emissions are due to the same electron population, we can fit radio and TeV data only with density and magnetic field values not easy to explain in that environment [36]. The only model that can well fit γ-ray data is a hadronic power-law distribution with an index $\gamma = 3.0 \pm 0.1$ and a cut-off at $E_{cut} = 6 \pm 1$ GeV, fixing a density $n = 100$ cm^{-3} and a magnetic field $B = 70$ μG, in agreement with the values predicted by acceleration theoretical models and with radio and TeV data (fig.2c)[36].

2.3. Discussion

W44 confirms that SNRs accelerate galactic CR protons in correspondence with shocks; they interact with the nuclei of the surrounding environment producing π^0 that decay in γ-ray photons

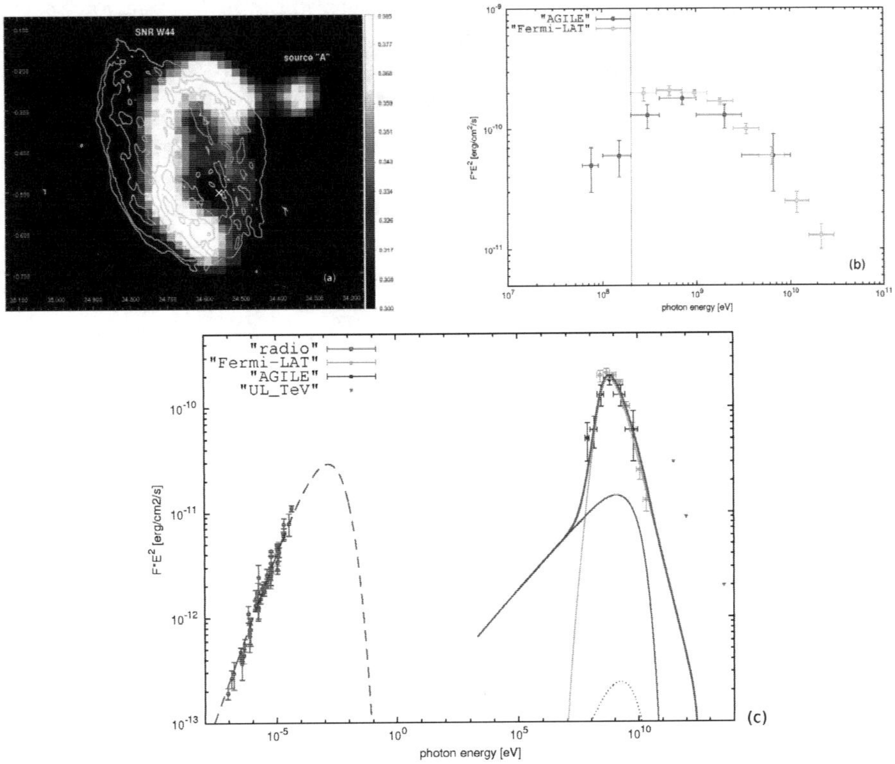

Figure 2: (a): W44 region AGILE γ-ray intensity map (Galactic coordinates), in the energy range 400 MeV-3 GeV obtained by integrating all available data collected during the period July 2007 and April 2011. Green contours show the 324 MHz radio continuum flux density detected by the Very Large Array. The white cross indicates the position of the pulsar PSR B1803+01. Source A does not appear to be associated with W44. **(b)**: combined AGILE (red) and Fermi/LAT (green) spectral energy distribution (SED) for SNR W44. AGILE points are in the range 50 MeV-10 GeV and Fermi/LAT data span the energy range 0.2 - 30 GeV [3]. **(c)**: theoretical modeling of the broad-band spectrum of SNR W44 superimposed with radio data (in red color), γ-ray data of image (b) (in blue color) and TeV upper limits (magenta triangles). This is an hadronic model characterized by B =70 μG and n =100 cm^{-3}. The yellow curve shows the neutral pion emission from the accelerated proton distribution discussed in the text. The green curves show the electron contribution by synchrotron (dashed curve), Bremsstrahlung (solid curve), and IC (dotted curve) emissions. The red curve shows the total gamma-ray emission.[36]

[39]. However the spectral index found both in AGILE and Fermi-LAT spectra, γ = 3.0, is steeper than the one predicted by linear and non-linear DSA models like the index of all other SNR spectra (section 1) . We need to consider other physical mechanisms that can modify the predicted spectrum in order to explain experimental data. In [43] Fermi-LAT spectrum is explained by the so called Alfvén damping: when accelerated ions scatter with neutrals (e.g. in a Molecular Cloud), Alfvén wave propagation is prevented by their friction in a well determined frequency band [40], producing a less efficient acceleration and a consequent steepening of the spectrum of exactly one unit [43].
The AGILE data seem to confirm this possibility but a new deeper analysis could show an even more complex scenario. W44 could be divided in two different regions: the main one, in correspondence with high density ionized gas and molecular cloud, that contributes to γ-ray emission in the range 400 MeV-1 GeV and the SouthWestern one that emits in the range 1-3 GeV. They could give different contributions to W44 spectrum because of different environments and,

consequently, different physical mechanisms for CR acceleration and propagation.
We are considering this possibility, taking into account a different behavior of diffusion coefficient in the two regions [32] and the Alfvén damping correlated to the presence of the molecular cloud [43]: we will show our results in a forthcoming paper [27].

3. Conclusions

The SNR W44 AGILE data confirmed the importance of this object in the CR acceleration context. For the first time AGILE obtained spectral point below $E = 200$ MeV giving the final proof of CR acceleration by SNRs. Our data confirm a spectral index steeper than the one predicted by theoretical models as Fermi-LAT found; this is not the first case for SNRs and it highlights the importance of deeply understanding SNR environment and physical mechanisms related to acceleration and propagation. New AGILE data analysis seems to show a more complex structure of W44 that could result in a composite spectral contribution coming from different SNR regions and consequently two different physical mechanisms.
Further developments of this hypothesis will be explained in a forthcoming paper [27].

References

[1] Abdo A. *et al.*, 2009,*ApJ Lett.*, **706**, L1L6.
[2] Abdo A. *et al.*, 2010, *ApJ Letters*, **710**, L92-L97
[3] Abdo A. *et al.*, 2010, *Science*, **327**, 1103-1106
[4] Abdo A. *et al.*, 2010, *ApJ*, **712**, 459-468
[5] Abdo A. *et al.*, 2010, *ApJ*, **718**, 348-356
[6] Abdo A. *et al.*, 2011, *ApJ*, **734**, 28-51.
[7] Acciari V. A. *et al.*, 2009, *ApJ*, **698**, 133-137
[8] Acciari V. A. *et al.*, 2010, *ApJ*, **714**, 163-169
[9] Acciari V. A. *et al.*, 2011, *ApJ*,**730L**, 720-725
[10] Aharonian F. A. *et al.*, 2001, *A&A*, **370**, 112-120
[11] Aharonian F. A., 2004, *Very high energy cosmic gamma radiation: a crucial window on the extreme Universe*, (Singapore: World Scientific Publishing)
[12] Aharonian F. A. *et al.*, 2007, *A&A*, **464**, 235-243
[13] Aharonian F. A. *et al.*, 2008, *A&A*, **481**, 401-410
[14] Aleksic J. *et al.*, 2012, *A&A*, **541**, 13-23
[15] Amato E. and Blasi P. , 2005, *MNRAS*, **364**, 76-80
[16] Amato E. and Blasi P. , 2006, *MNRAS*, **371**, 1251-1258
[17] Atoyan A. M. *et al.*, 2000, *A&A*, **354**, 915-932
[18] Atoyan A. M. *et al.*, 2000, *A&A*, **355**, 211-220
[19] Atoyan A. M. & Dermer C D, 2011, arXiv:1111.4175
[20] Bell A. R., 1978, *MANRAS*, **182**, 147-156
[21] Bell A. R., 1978, *MANRAS*, **182**, 443-455
[22] Berezhko E. G. & Voelk H J, 2007, *Astrophys. Journal*, **661**, L175-178.
[23] Berezinskii V. S. *et al.*, 1990, *Astrophysics of Cosmic Rays*, (Amsterdam: North-Holland)
[24] Blandford R. and Eichler D., 1987, *Physics Reports*, **154**, 1-75
[25] Butt Y., 2009, *Nature*, **460**, 701-704.
[26] Caprioli D., 2011, *JCAP*, **05**, 026-075
[27] Cardillo M. *et al.*, in preparation
[28] Castelletti G. *et al.*, 2007, *A&A*, **471**, 537-549
[29] Clark D. H. & Caswell J L, 1976, Mon.Not.R.astr.Soc., **174**, 267-305
[30] Claussen M. J., Frail D. A. & Goss W. M., 1997, *ApJ*, **489**, 143-159
[31] Fermi E., 1949, *Physical Review*, **75**, 1169-1174
[32] Gabici S., *et al.*, 2009, *MNRAS*, **396**, 1629-1639
[33] Giacani E. B. *et al.*, 1997, *AJ*, *113*, 1379-1390.
[34] Ginzburg V. L. & Syrovatskii S I, 1964, *The Origin of Cosmic-rays*, (Pergamon Press)
[35] Giordano F. *et al.*, 2012, *ApJLetters*, **744**,L2-L6
[36] Giuliani A., Cardillo M., Tavani M. *et al.*, 2011, *ApJLetters*, **742**, L30-L35
[37] Giuliani A. *et al.*, 2010, *A&A*, **516**, L11-L14
[38] Hoffman I. M. *et al.*, 2005, *ApJ*,**627**, 803-812
[39] Kelner, S. R., Aharonian, F. A., & Bugayov, V. V. 2006, *PhRvD*,**74**, 034018
[40] Kulsrud R & Pearce W. P., 1969, *ApJ*, **156**, 445-469
[41] Malkov M. A. & Drury L. O'C., 2001, *RPPh*, **64**, 429-481
[42] Malkov M. A. & Diamond P. H., 2006, *ApJ*, **642**, 244-259
[43] Malkov M. A., Diamond P. H. & Sagdeev R. Z., 2011, *Nature Communication*, **2**, 194.
[44] Reach W. T. *et al.*, 2005, *AJ*, **618**, 297-320
[45] Rho J. *et al.*, 1994, *AJ*, **430**, 757-773
[46] Seta M. *et al.*, 2004, *AJ*, **127**, 1098-1116
[47] Tavani M. *et al.*, 2010, *ApJ*, **710**, L151-L155
[48] Tidman D. A., 1966, *ApJ*, **144**, 615-627
[49] Torres D. F., Romero G. E., Dame T. M., Combi J. A. & Butt Y. M., 2003, *Physics Reports* **382**, 303-380.
[50] Uchiyama Y., Blandford R. D., Funk S., Tajima H. & Tanaka T., 2010, *ApJ*, **723**, L122-L126
[51] Watson M. G. *et al.*, 1983, *IAUS*, **101**, 273W
[52] Wolszczan A. *et al.*, 1991, AJ, **372**, L99-L102
[53] Wootten H. A., 1977, *ApJ*,**216**, 440-445
[54] Zweibel E. G. & Shull J. M., 1982, *ApJ*,**259**, 859-868

The Classical Cepheid Distance scale in the NIR bands

L. Inno[1], N. Matsunaga[2], G. Bono[1,3], F. Caputo[3], C. D. Laney[4], A. Pietrinferni[5], F. Primas[6], M. Romaniello[6]

[1] Dipartimento di Fisica, Università di Roma Tor Vergata via della Ricerca Scientifica 1, 00133 Rome, Italy; laura.inno@roma2.infn.it
[2] Kiso Observatory, Institute of Astronomy, School of Science, The University of Tokyo 10762-30, Mitake, Kiso-machi, Kiso-gun,3 Nagano 97-0101, Japan
[3] INAF–OAR, via Frascati 33, Monte Porzio Catone, Rome, Italy
[4] Dept. of Physics and Astronomy, N283 ESC, Brigham Young University, Provo, UT 84601, USA South African Astronomical Observatory, P.O. Box 9, Observatory 7935, South Africa
[5] Osservatorio Astronomico di Teramo, Via Mentore Maggini s.n.c., I-64100 Teramo, Italy
[6] European Southern Observatory, Karl-Schwarzschild-Str. 2, 85748 Garching bei Munchen, Germany

E-mail: `laura.inno@roma2.infn.it`

Abstract. The Period–Luminosity (PL) relation of classical Cepheids is one of the most popular standard candle, being the first rung of the cosmic distance ladder. We discuss both theoretical and empirical features of the PL, Period–Luminosity–Color (PLC) and Period–Wesenheit (PW) relations using the largest optical (V,I; OGLEIII) and near-infrared (NIR, J,H,K_S; IRSF/SIRIUS) data sets ever collected for classical Cepheids in the Magellanic Clouds (MCs). The MCs are indeed a fundamental benchmark to calibrate the cosmic distance scale. We found that the PW relations are linear over the entire period range and show a very low dispersion. Moreover, their slopes appear independent of the metallicity since they attain similar values in the MCs and in the Galaxy. This indicates that Cepheids in NIR bands obey to universal PW relations.

1. Introduction

Classical Cepheids play a key role in constraining the cosmic distance scale. They are yellow supergiant variable stars, they can be observed at great distances and they obey to a well known Period–Luminosity (PL) relations. Since its discovery by Miss Henrietta Leavitt in the 1912, the PL relation has played a pivotal role in constraining Galactic and extra-galactic distances, leading to the discovery of the expansion of the Universe by Edwin Hubble. However, after one century the universality of the Cepheid PL relations is still under debate. This means the possibility to use the Galaxy-calibrated Cepheid PL relations to estimate the distance of any galaxy containing Cepheids, independent of its chemical composition. During the last ten years theoretical and observational efforts were devoted to investigate the dependence of the Cepheid pulsation properties on metallicity. Indeed, a metallicity dependence could introduce systematic errors in the evaluation of extragalactic distances, and in turn on the Hubble constant, H_0 [3].

In this investigation, we briefly summarize some of the main problems affecting the Cepheid distance scale in the near-infrared (NIR, J,H,K_S) bands. In particular, we analyze the PL

Journal of Physics: Conference Series **383** (2012) 012008 doi:10.1088/1742-6596/383/1/012008

relations in the NIR bands for the Cepheids of the Magellanic Clouds (MCs), using the largest NIR data set ever collected.

2. The PL and PLC relations

At the basis of all Cepheid distance determinations lies a Period–Luminosity–Color (PLC) relation:

$$M_\lambda = \alpha_\lambda + \beta_\lambda \ LogP + \gamma_\lambda (CI)_0 \quad (1)$$

where M_λ is the magnitude in the λ photometric band (i.e. $\lambda = B, V, I, J, H, K_S, ...$) and $(CI)_0$ is the mean intrinsic color index. Traditionally, the relation widely used to estimate Cepheid distances is the PL relation found by H. Leavitt: $M_\lambda = \alpha_\lambda + \beta_\lambda \log P$. This relation is the projection on the Period–Magnitude plane of the correct PLC relation. Since only the PLC relation can account for the Cepheid pulsational properties, the PL relation is expected to have an intrinsic dispersion. The PLC relations is a simply restatement of the Stefan's law that is applicable to stars on individual basis. The Cepheid instability strip defines a range of luminosity, colors and periods over which pulsation is a stable mode for the star and is therefore an observable: these constraints control the statistical properties of the ensemble of Cepheid variable stars. Recent theoretical predictions based on nonlinear, convective Cepheid models indicate that the instability strip boundaries in the Period-Magnitude planes are better described by quadratic P–L relations [1]. If we look at the Fig. 1 we note that at each given luminosity the pulsation instability occurs over a finite range of effective temperatures. This finite width of the instability strip and particular distributions of Cepheids within the instability strip cause indeed an intrinsic scatter in any PL relation.

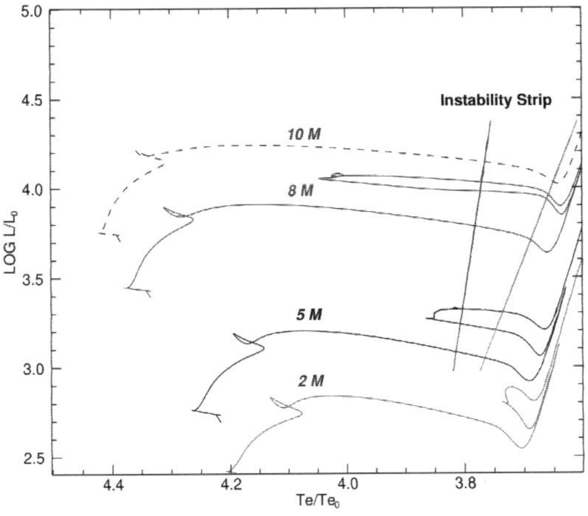

Figure 1. The evolutionary track for stars with different masses. The red and the blue boundaries of the instability strip, according to [1], are also overplotted (blue and red solid lines). The tracks of the intermediate ($5M_\odot$ and $8M_\odot$) mass stars cross the instability strip three times and the $10M_\odot$ stars crosses it once, while the track of low mass star ($2M_\odot$) do not cross the strip at all.

Moreover, when we consider the observable quantities, we have to take into account the effect of interstellar reddening and absorption. It's difficult to separate the effects of extinction from the effect of metallicity in equation 1 (this is the so called: metallicity–reddening degeneracy). However, the debate on the role played by the chemical composition on the pulsational properties of Cepheids is still open, with different theoretical models and observational results leading to markedly different conclusions. The atmospheres of stars like Cepheids, having effective temperatures typical of G and K supergiants, are affected by changes in the atmospheric metal abundance (these changes in chemical composition produce also a change in the overall stellar structure). Thus, it is expected that the colors and magnitudes of Cepheids, and their corresponding PL relations, should be a function of metallicity. At each given luminosity an increase in the metal content shifts the instability strip toward cooler effective temperatures. This implies that metal-rich pulsators have on average longer periods than metal-poor ones with the same mass and luminosity. It turns out that the mean bolometric magnitude, at fixed period and for an uniformly populated instability strip, increases at larger metal contents. Owing to the dependence of the bolometric correction on the effective temperature, this effect is further enhanced when visual magnitudes are taken into account.

2.1. The NIR bands

Recent theoretical models predict that NIR PL relations are almost completely unaffected by metallicity and show also a smaller dispersion due to the reduced width of the instability strip [1]. These models predict a metallicity dependence of the PL relations in V and I, but empirical analyses of the metallicity corrections to Cepheids distances have not yet achieved a firm result. Moreover, reddening and absorption are much smaller at longer wavelengths. At the J,H, K_S wavelengths, interstellar reddening and absorption, and the sensitivity to heavy-element abundance are all much smaller. We also have to observe that, as a results of the physical proprieties of Cepheids, the PL Relations in the infrared show less scatter because the width in temperature of the instability strip is narrower. Moreover, the pulsation amplitude of a single Cepheid decreases for increasing wavelength. These means that we can approximate the mean NIR magnitude with the single epoch measurement taken at a given pulsation phase. A huge amount of data collected by the automatic surveys become available in the NIR bands, and the use of these single epoch measurements leads to a larger statistical significancy of the PL relations.

3. The PW relations

The Wesenheit indices was introduced by [4], who define a Wesenheit magnitude, W, by:

$$W(V,I) = V - R_{VI} \times (V - I); \tag{2}$$

where the ratio of total-to-selective absorption $R_{VI} = A_V/E(V - I)$ is adopted by [2], A_V is the absorption and $E(V - I)$ is the color excess or reddening. By construction this Wesenheit parameter is reddening-free:

$$W(V,I) = V_0 + A_V - R_{VI} \times (V - I)_0 - R_{VI} \times E(V - I) =$$

$$= V_0 - R_{VI} \times (V - I)_0 + A_{VI} - R_{VI} \times E(V - I)$$

where $A_{VI} = R_{VI} \times E(V - I)$ in the last term, so $W(VI,) \equiv W_0(V,I)$, which is the de-reddened W magnitude. Thus, W, for any given star, is dimmed only by distance and it is unaffected by extinction by its definition, but only to the degree that R_{VI} is known and is universal. Similarly any Wesenheit parameter based on an apparent magnitude, color, and a given reddening law can be computed for any combination of optical/near-IR bandpasses. The advantage of using these parameters is that uncertainties in the amount of extinction toward any given Cepheid, whether in the Magellanic Clouds or within the set of Galactic calibrators, is avoided once that the reddening law, which controls R_V and the analogous ratios for other Wesenheit parameters ($R_{\lambda_1\lambda_2}$), is chosen. Cepheids obey to the Period-Wesenheit (PW) relation:

$$W(\lambda_1, \lambda_2) = \alpha_{\lambda_1,\lambda_2} + \beta_{\lambda_1,\lambda_2}\ LogP \tag{3}$$

whereλ_1, λ_2 are any pair of photometric band. Theoretical predictions [1] show that these relations are quite tight, and show several additional advantages: they are almost independent of the pulsator distribution within the instability strip; they are linear over the entire period range; the effects of the mixing-length – l/Hp-- at fixed metal (Z) and Helium content (Y) on both the slope and zero-point are negligible; the adopted Y , at fixed Z and mass-luminosity relation, only affects the zero–point of the relations, while their slopes turn out to be almost independent of the chemical composition.

4. The PL and PW relations for the MCs

For this work we have used the OGLE - III Catalog of Variable Stars (CVS) and the IRSF/SIRIUS Catalog. The OGLE-III CVS [8] is a free catalog that can be downloaded at *http://ogledb.astrouw.edu.pl/ ogle/CVS/.* The InfraRed Survey Facility (IRSF) is located at the Sutherland observing station of the South African Astronomical Observatory (SAAO). It carried out an imaging survey of the MCs in the NIR bands J (1.25 μm), H (1.65 μm) and K_S (2.14 μm). The single–epoch J,H,K_S magnitudes for the OGLE-III FU Cepheids were extracted from the IRSF/SIRIUS Survey by [5]. We have then collected a sample of 3042 for the Large Magellanic Cloud (LMC) and 4150 for the Small Magellanic Clouds (SMC) Cepheids with V,I,J,H,K_S single–epoch measurements (see Fig. 2,Fig. 3 and Fig. 4). Using this dataset we can analyze the main features of the PL and PW relations. The data showed in Fig. 2 compare the PL(V) and PW(V,I) relations in the Optical bands for the LMC. According to the theoretical predictions discussed in the above section, the PW relation in Fig. 2 is tight and linear over the entire range of period. When we move to the NIR bands, we find that the PL is narrower [9]. The data showed in Fig. 3 compare the PL(J) and PW(J,K_S) relations in the NIR bands for the LMC. The dispersion around the PL(J) relation (left panel) is still large, due to the reddening effects, while the PW(J,K) (right panel) relation is very tight, as expected. The comparison between the PL(J) and PW(J,K_S) relations in the NIR bands for the SMC is given in Fig. 4. The dispersion around the PL(H) relation (left panel) and the PW(H,K_S) relation (right panel) are quite similar (a detailed discussion on the PW relations for MCs is given in Inno et al.,2012, in preparation). In the case of the PL(H) relations the dispersion can be related to the reddening, while for the PW relations, it is due to the intrinsic feature of the SMC, since the galaxy is elongated along the line of sight [10]. According to theoretical predictions, we can assume then that the dispersion around the PW relation is due only to the individual distances of Cepheids. The PW relation can than be used as a tool to investigate the geometry and the three dimensional structure of the MCs.

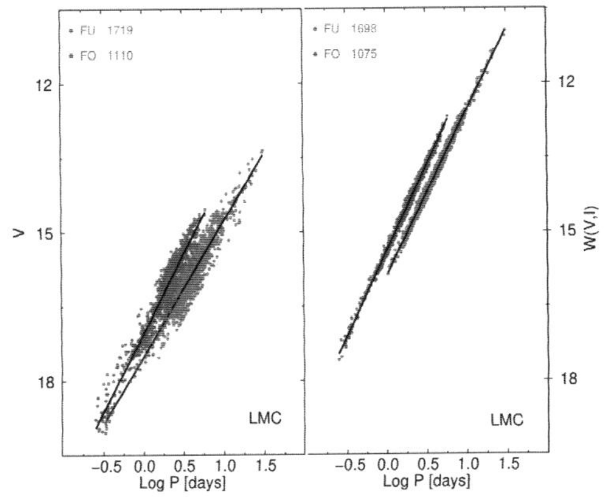

Figure 2. Comparison between the PL(V) (left panel) and PW(V,I) (right panel) relations for the LMC. Red dots display OGLE-III mean magnitudes for FU pulsators and green dots are the mean magnitude for FO pulsators.The solid lines show the linear fits. The dispersion around the PL is large and also non linear effect are evident, while the PW(V,I) relation shows linearity and also a very low dispersion.

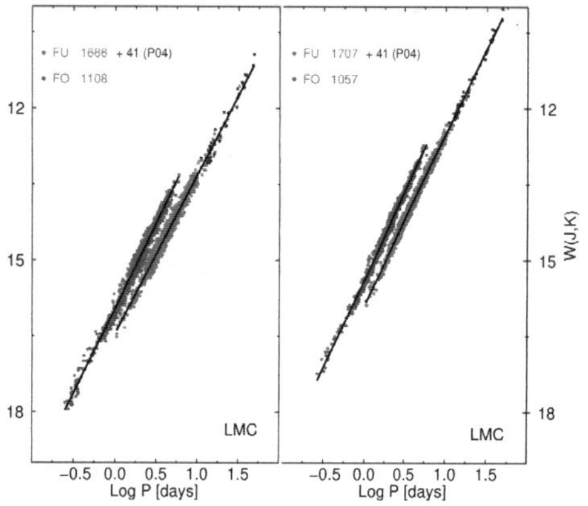

Figure 3. Comparison between the PL(J) and PW(J,K_S) relations for the LMC. Red dots display IRSF/SIRIUS mean magnitudes for FU pulsators, while purple dots are the mean magnitude for 41 Cepheids by P04. The green dots are the mean magnitude for FO pulsators. The solid lines show the linear fits. The dispersion around the PL relation (left panel) is still large, due to the reddening effects, while the PW(J,K_S) (right panel) relation is very tight

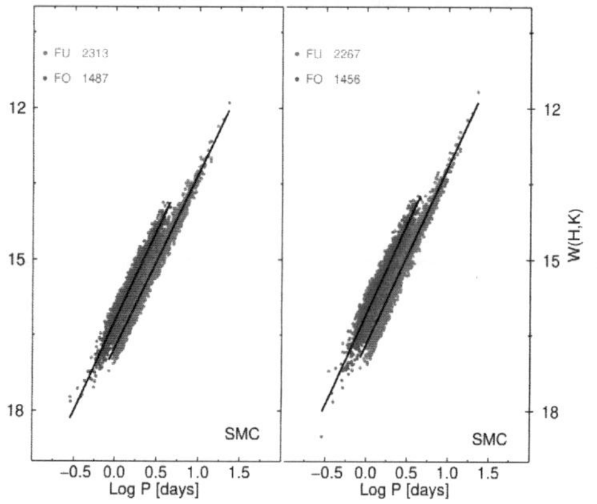

Figure 4. Comparison between the PL(H) and PW(H,K_S) relations for the SMC. Red dots display IRSF/SIRIUS mean magnitudes for FU pulsators, and the green dots are the mean magnitude for FO pulsators. The solid lines show the linear fits. In this case we find that the dispersion around the PL(H) relation (left panel) and the PW(H,K_S) relation (right panel) are quite similar. This is due to the intrinsic proprieties of the SMC.

5. Conclusions

The main aim of this work is to investigate the properties of the Cepheids as distance indicators. Classical Cepheids play a key role in distance determinations, and the Magellanic Clouds are the fundamental benchmark to calibrate the astronomic distance scale. In particular we focused on the PL relations for the MC Cepheids and we find that, when we move from Optical (V) to NIR bands, the dispersion due to non-linearity effects is lower,according to theoretical models. Moreover, metallicity effects on the PL relation are expected to decrease in amplitude with increasing wavelength. Although the extinction term is also reduced in the NIR bands, we need to correct the PL for reddening effects. Another approach is to use the Wesenheit functions. In studying the PW relations for MCs Cepheids we find that they are linear over the entire range of the period, as predicted by theoretical models. Moreover we find a very low statistic dispersion around the PW relation, suggesting that the Wesenheit function can be safely used to derive individual Cepheid distances (see also [6]). The last and most important finding is that the PW relations in both the MCs are very similar, supporting the theoretical prediction that the slope of the PW relations is metallicity independent. Current finding thus suggest the possibility of a universal PW relation for Cepheids.

References

[1] Bono, G., Caputo, F., Marconi, M., & Musella, I. 2010, *Astrophysical Journal*, **715**, *277*
[2] Cardelli, J. A., Clayton, G. C., & Mathis, J. S. 1989, *Interstellar Dust, 135, 5P*
[3] Freedman, W. L., & Madore, B. F. 2010, *Annual Review of Astronomy and Astrophysics*, **48**, *673*
[4] Madore, B. F. 1982, *Astrophysical Journal*, **253**, *575*
[5] Matsunaga, N., Feast, M. W., & Soszyński, I. 2011,*Monthly Notices of the Royal Astronomical Society*, **413**, *223*

Journal of Physics: Conference Series **383** (2012) 012008 doi:10.1088/1742-6596/383/1/012008

[6] Ngeow, C.-C. 2012, *Astrophysical Journal*, **747**, *50f*
[7] Persson, S. E., Madore, B. F., Krzemiński, W., et al. 2004, *Astronomical Journal*, **128**, *2239* (P04)
[8] Soszyńnski, I., Poleski, R., Udalski, A., et al. 2008, *Acta Astronomica*, **58**, *163*
[9] Storm, J., Gieren, W., Fouqué, P., et al. 2011,*Astronomy & Astrophysics*, **534**, *A95*
[10] van den Bergh, S. 2000, *The galaxies of the Local Group, by Sidney Van den Bergh.* Published by Cambridge, UK: Cambridge University Press, pp.*92-162*

Young Researcher Meeting in Rome 2012 IOP Publishing
Journal of Physics: Conference Series **383** (2012) 012009 doi:10.1088/1742-6596/383/1/012009

On the kinematic structure of the Carina dwarf spheroidal galaxy

Michele Fabrizio[1], Ivan Ferraro[2], Giacinto Iannicola[2], Giuseppe Bono[1,2], Mario Nonino[3] and Frédéric Thévenin[4]

[1]Dept. Physics, Univ. Rome "Tor Vergata", Via della Ricerca Scientifica, 1 – 00133, Rome
[2]INAF–OAR, via Frascati 33, Monte Porzio Catone, Rome, Italy
[3]INAF–OAT, via G.B. Tiepolo 11, 40131 Trieste, Italy
[4]Univ. Nice Sophia-antipolis, CNRS, OCA, Lab. Lagrange, BP 4229, 06304 Nice, France

E-mail: michele.fabrizio@roma2.infn.it

Abstract. We present a semi-automatic interactive code, ROma Stellar Spectroscopic Analyzer (ROSSA), to measure radial velocities (RVs) from stellar spectra, covering a broad range of spectral resolutions. The measurement is supported by a graphical interface to change the fitting function (Gaussian or Moffat) and their parameters such as, e.g., symmetry and wings contribution. ROSSA has been applied to the kinematical structure of the Carina dwarf spheroidal galaxy, in order to obtain homogeneous RV measurements of ~1 400 stars collected with various VLT spectrographs (FORS2, FLAMES/GIRAFFE medium- and high-resolution grisms). The spectroscopic targets cover different evolutionary phases and stellar populations of Carina, from the red giant branch to phases of old (horizontal branch) and intermediate age (red clump) helium burning stars. These stars are the tracers of two different star formation episodes (2-6 and 12 Gyr).

1. Introduction

Dwarf galaxies are fundamental laboratories to investigate the influence of the environment on star formation and chemical evolution in stellar systems that are several order of magnitudes smaller than giant galaxies. Empirical evidence indicates that in the Local Group (LG) together with the dwarf ellipticals (dE, M32-like) and the dwarf spirals (dS, M33-like) we are facing three different dwarf morphological types. The dwarf spheroidals (dSphs) show either single (Cetus, [15]; Tucana, [16]) or multiple star formation events (Carina, [2]), are spheroidal in morphology and lack of neutral hydrogen [3]. The dwarf irregulars (dIs) host a mix of old and young stellar populations, but they have recently experienced an intense star formation episode [17, 4], moreover, they have irregular morphology and host a significant fraction of gas and molecular clouds. The transition dwarfs have properties between the dSphs and the dIs. The observational scenario concerning the dwarfs in the LG was recently enriched by the discovery of ultra-faint dSphs in the SDSS [21, 18, 20], but their properties still need to be investigated in detail. Although the morphological classification appears robust, we still lack firm quantitative constraints concerning the evolution/transition between different dwarf morphological types. Kormendy et al. [12, 13] found a well defined difference between ellipticals (Es) and spheroidals [14]. Early type galaxies are distributed along a sequence moving from cD to dEs, while spheroidals are distributed along a sequence that overlaps with spirals and dwarf irregulars (see Fig. 1 in [14]). This empirical dichotomy seems to suggest that E and Sph galaxies are stellar systems that underwent different formation and evolution processes. In particular, the Sphs might be either spirals or irregulars that lost their gas or transformed it into stars. However, there is no general consensus concerning the E–Sph dichotomy, since the correlation between the shape of the brightness profile and the galaxy luminosity is continuous when moving from Es to Sphs [9, 8, 7].

Precise and homogeneous photometric and spectroscopic data are required to address the above open issues concerning the formation and evolution of these systems. The advent of multi-object spectrographs at the 8-10m class telescopes provided the unique opportunity to collect sizable samples of low-, medium- and high-resolution spectra covering the entire body of nearby dwarf galaxies. By using tens of radial velocity (RV) measurements, it has been suggested by [11] that Sextans dSph hosts kinematic substructures, in particular, they found that the RV dispersion of the stellar population located across the center of this system is close to zero and increases outside the core. A different kinematic status has also been suggested by [1], using ~500 red giant (RG) stars, for the two distinct stellar populations in the Sculptor dSph.

In this context, Carina plays a key role because it is relatively close (μ=20.10 mag), shows at least two well separated SF episodes (t=2-6 and 12 Gyr) and it is metal-poor ($<$[Fe/H]$>\sim -1.7$ dex). In Fabrizio et al. [5] we provided RV measurements for ~1 400 Carina candidate stars showing the occurrence of secondary kinematical feature across the center of the galaxy, suggesting a correlation with the transition properties detected in dwarf ellipticals [10]. The measurement of RVs was performed adopting an interactive code (ROSSA) described in the following.

2. ROma Stellar Spectroscopic Analyzer

The f77-code ROSSA is a semi-automatic software, with a graphical interface, meant to obtain accurate measurements of stellar radial velocities, independently from the evolutionary phase or metallicity of the star or the resolution of spectra.

To measure the RVs, we start measuring the peaks of well known spectral features. In particular, we selected strong spectral lines easily to recognize by eye inspection in observed spectra (e.g. H_β[1], Mg I triplet[2], near infrared Ca I triplet[3], Fe I[4]). All spectra are normalized to continuum, that is estimated with an iterative smoothing algorithm in order to avoid an over or underestimate across the strongest lines. The fitting of single or multiple blending lines is performed by using a Gaussian or a Moffat function. The user could fit each line by a Gaussian or by a Moffat function with different β values, to properly account the contribution of wings in the line fit. These peak positions give a preliminary estimation of RV and are weighted by the inverse of their amplitudes (a sharp line has more weight than a broad one). The preliminary estimate of the RV is validated by visual inspection and, when judged to be satisfactory, it was adopted to perform an automatic estimate of the RV.

The user must provide a line list with the reference wavelengths of several spectral lines (single or blended) that appear in the observed spectrum. Each line is fitted with a Moffat function ($\beta = 2$) that has a peak value, converted in RV, inside the spread of RV measured by hand in the previous step. The error in the RV based on individual lines was assumed to be equal to the sigma of the fitting function. In particular, we assume $\epsilon_{RV} = c\sigma_\lambda/\lambda$ km s^{-1}. Several checks are performed to discard spurious line identifications and fittings. In particular, are neglected lines with an amplitude smaller than half or greater than twice the line width given by the spectrograph resolution ($\sigma_\lambda = \lambda/R$). The RV of the entire spectrum was estimated as

[1] H_β: $\lambda = 4861.34$ Å
[2] MgT: $\lambda = 5167.33, 5172.70, 5183.62$ Å
[3] CaT: $\lambda = 8498.02, 8542.09, 8662.12$ Å
[4] Fe I: $\lambda = 5429.70, 5434.53, 5446.92, 5455.62$ Å

a weighted mean over the different fitted lines (weights are chosen as the inverse of the errors). The error in the RV of the entire spectrum is the error of the weighted mean.

In Fig. 1 are shown two red giant stars (RG) of Carina dSph in a range of ≈2 mag. From top

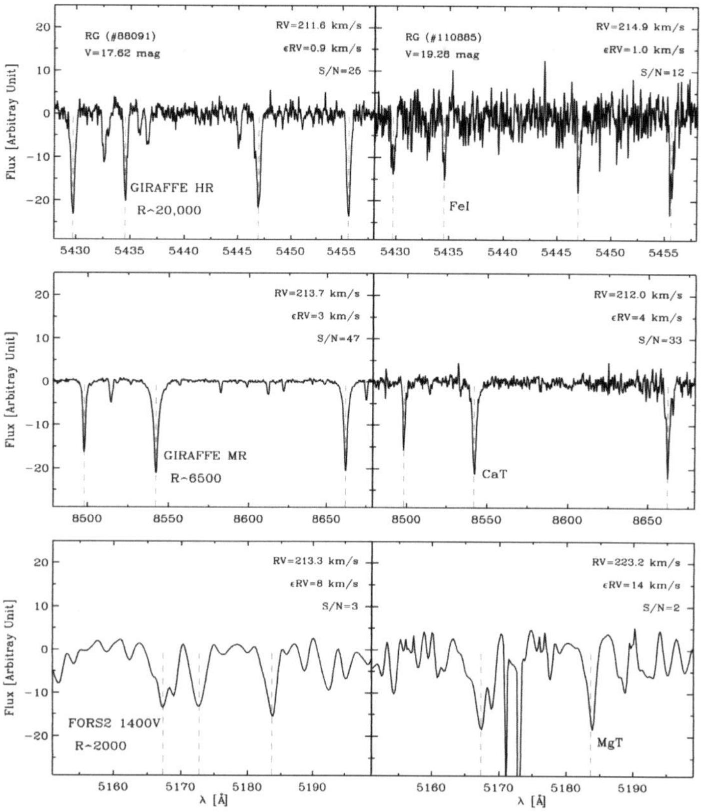

Figure 1. Left panels: from top to bottom are shown spectra of a Carina RG star with $V = 17.62$ mag, collected with three different spectrographs with decreasing spectral-resolution. Right panels: same as the left panels, but for a fainter Carina RG star with $V = 19.28$ mag.

to bottom panels, are shown the same stars observed with different spectrograph at ESO/VLT: FLAMES/GIRAFFE HR10 grism ($R \sim 20\,000$), FLAMES/GIRAFFE LR08 grism ($R \sim 6\,500$) and FORS2 1400V grism ($R \sim 2\,000$), respectively. Dotted lines display the spectral features adopted for the visual preliminary RV determination. The signal-to-noise ratios (S/N), RV values with their uncertainties are also labelled. All RVs agree, within the error, quite well. The uncertainties are bigger for spectra with lower S/N.

The power of this algorithm is the versatility across a broad range of spectrograph and stellar parameters. Indeed, the pros with respect to the most adopted method of the cross-correlation with a theoretical or empirical template, is the absolute independence from the stellar parameters. Therefore, it is possible to measure RV with high accuracy, for any metallicity, gravity and effective temperature. The only constraint is the visibility of spectral lines. The accuracy of the result, of course, will depend on the number of lines measured. Therefore, for example in an hot horizontal branch star ($T_{eff} \sim 20\,000$ K) will be measured just few broad lines, like H_β and He I, obtaining an accuracy smaller (~10 km s^{-1}) than one obtained with a red giant star with more than one hundred measured lines (~2 km s^{-1}).

Moreover, in order to improve the accuracy of the RV measurements, the user could choose a modified Moffat function, based on two different β-values, to fit the asymmetric lines[5] [19]. Fig. 2 shows an example of the asymmetric fit for one line of the Mg I triplet, at $\lambda = 5183.62$ Å, on a FORS2 low-resolution spectrum. This Mg I line, at the resolution of ~2 000, appears to be

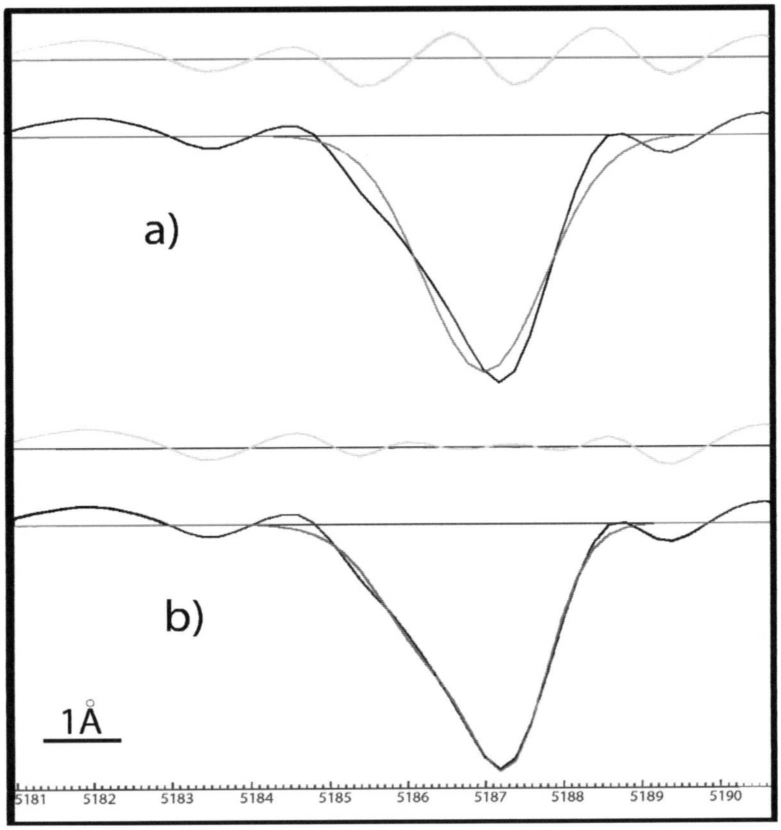

Figure 2. Fits with a Moffat function of the Mg I line on a low-resolution spectrum (not RV corrected) collected with FORS2@VLT (bottom right panel of Fig. 1). The fit showed in the top panel was performed by using a symmetric function, while the fit of the bottom panel was performed by adopting an asymmetric Moffat function with two different β-values. The red curves display the fit. The green curves are the residuals plot.

blended with the near absorption features (MgH, Fe I and Ni I), therefore shows a pronounced asymmetry. The green lines display the residuals of the fitting procedure, showing the better removal of the line wings (panel b) in both sides of the line respect to a symmetric fit with a Gaussian (panel a).

In order to simplify the measurement on the spectra, we introduced a list of possible options. For example, there is a complete list of options to produce a fit with fixed/variable peak or width or continuum level. A screenshot of the interactive software is shown in Fig. 3.

3. Conclusions

We developed a new software (ROSSA) to perform accurate radial velocity measurements from spectra covering a broad range of spectral resolutions. The main advantage of ROSSA, when compared with similar codes based on the cross-correlation technique, is that no theoretical or empirical grid of stellar templates is required. This means that ROSSA is particularly useful in dealing with large samples of stellar spectra collected with different spectrographs and covering a broad range of signal-to-noise ratios. At the same time, ROSSA can easily analyze spectra of stars in different evolutionary phases, and therefore covering a broad range of effective temperatures and surface gravities. We validated ROSSA by investigating the

[5] The details of the modified Moffat fitting procedure will be described in a forthcoming paper, Ferraro et al. 2012.

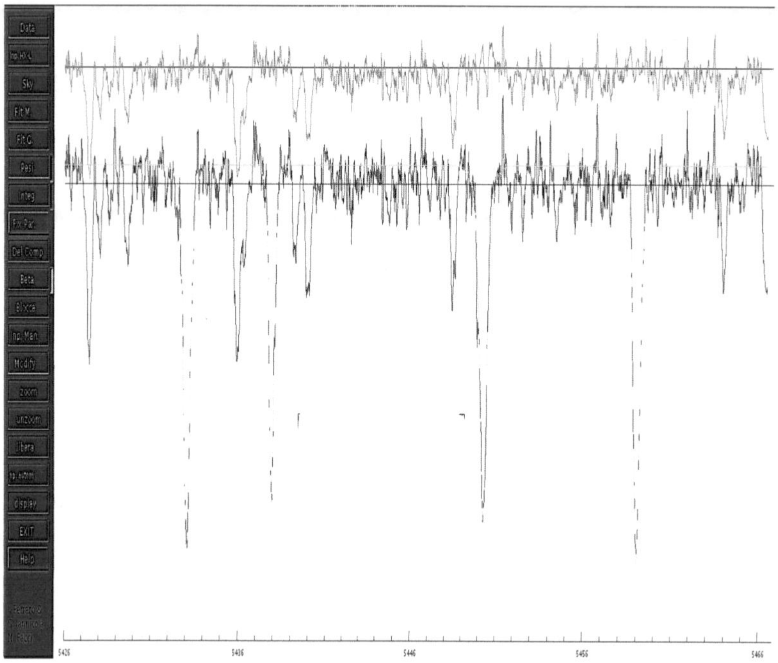

Figure 3. A screenshot taken from ROSSA on the FLAMES/GIRAFFE HR spectrum (top-right panel in Fig. 1). ROSSA was applied to four Fe I lines. On the left is visible a list of boxes listing the options that the user can use interactively to improve the quality of the fit. The color coding is the same as in Fig. 2.

kinematical structure of the Carina dSph. We used ∼1 400 spectra covering the entire body of the galaxy and we found evidence of substructure(s) across the central regions [5, 6].

Acknowledgments
MF would like to thank the organizers for the stimulating meeting. This work has been done in collaboration with the Carina project: M. Monelli, P. B. Stetson, A. R. Walker, P. François, R. Buonanno, F. Caputo, C. E. Corsi, M. Dall'Ora, R. Gilmozzi, C.R. James, T. Merle, L. Pulone and M. Romaniello.

References

[1] Battaglia G., Helmi A., Tolstoy E., Irwin M., Hill V. and Jablonka P. 2008, *ApJ*, **681**, L13
[2] Bono G., et al 2010, *PASP*, **122**, 651
[3] Bouchard A., Jerjen H., Da Costa G.S. and Ott J. 2005, *AJ*, **130**, 2058
[4] Cole A.A. 2010, *PASA*, **27**, 234
[5] Fabrizio M., et al 2011, *PASP*, **123**, 384
[6] Fabrizio M., 2011, PhD Thesis, Univ. Rome "Tor Vergata"
[7] Ferrarese L., et al. 2006, *ApJS*, **164**, 334
[8] Gavazzi G., Donati A., Cucciati O., Sabatini S., Boselli A., Davies J. and Zibetti S. 2005, *A&A*, **430**, 411
[9] Jerjen H. and Binggeli B. 1997, *in ASP Conf. Ser.* **116**, The Second Stromlo Symposium: The Nature of Elliptical Galaxies (San Francisco: ASP), 239
[10] Jerjen H., Kalnajs A. and Binggeli B. 2000, *A&A*, **358**, 845
[11] Kleyna J.T., Wilkinson M.I., Evans N.W. and Gilmore, G. 2004, *MNRAS*, **354**, L66
[12] Kormendy J. 1985, *ApJ*, **292**, L9
[13] Kormendy J. 1987, *in IAU Symp.* **127**, Structure and Dynamics of Elliptical Galaxies (Cambridge: Cambridge Univ. Press), 17
[14] Kormendy J., Fisher D.B., Cornell M.E. and Bender R. 2009, *ApJS*, **182**, 216
[15] Monelli M., et al. 2010, *ApJ*, **720**, 1225
[16] Monelli M., et al. 2010, *ApJ*, **722**, 1864
[17] Sanna N., et al. 2009, *ApJ*, **699**, L84
[18] Tolstoy E., Hill V., and Tosi M. 2009, *ARA&A*, **47**, 371
[19] Wallerstein G., et al. 1992, *MNRAS*, **259**, 474

[20] Wyse R.F.G. 2010, *AN*, **331**, 526
[21] Zucker D.B., et al. 2006, *ApJ*, **650**, L41

Young Researcher Meeting in Rome 2012 IOP Publishing
Journal of Physics: Conference Series **383** (2012) 012010 doi:10.1088/1742-6596/383/1/012010

Optical identifications of celestial high energy sources with the Telescopio Nazionale Galileo

Sara Turriziani

Dipartimento di Fisica, Universit di Roma "Tor Vergata", Via della Ricerca Scientifica 1, 00133, Roma, Italy

E-mail: sara.turriziani@roma2.infn.it

Abstract.
To ascertain the nature of celestial high energy sources, it is crucial to identify their optical counterparts. However, the currently available astronomical public optical databases do not provide an adequate support for a systematic high energy sources identification work. In particular, the optical limiting magnitude represents a severe limitation since the deepest flux limits reached by X-ray surveys require of course similarly deeper optical catalogs to homogeneously sample the available parameter space. Nonetheless, dedicated spectroscopic campaigns are being carried out successfully with the Telescopio Nazionale Galileo (TNG), a 4-m class telescope. To set up a winning observational campaign, the first and most important step is to define a strong science case, as it will allow for selections of good targets for observations: the key is to increase the identification efficiency while keeping down the required telescope time. In this context, as the Principal Investigator, I will give an overview of the first spectroscopic campaign carried out at the TNG to identify Swift X-ray serendipitous sources, and I will show the valuable results achieved with only one night of observations. As a second example, I will review the strategy for the northern-sky classification of candidate blazars associated to unidentified *Fermi* γ-ray sources, and I will show the results coming from the related observational campaign at TNG I have been involved during the last two years.

1. Matter of interest

"The problem of optical identification of high energy (X- and γ-ray) sources is a classic in modern astronomy. It is only through the optical studies that one can gain complete understanding of objects [..]" [1].

Currently available astronomical public optical databases do not provide an adequate support for a systematic high energy sources identification work [see e.g. 2]. The main issues to solve are essentially related to the poor positional accuracy of high energy sources coordinates with respect to optical ones; on the other hand, the limiting magnitude of available optical surveys can represents a severe limitation as the deepest flux limits reached by current high energy missions requires of course similarly deeper optical catalogs to homogeneously sample the parameter space.

To date, the deepest wide optical survey, the Sloan Digital Sky Survey [SDSS; 3], has a limiting magnitude of $r' \sim 22$. Much deeper serendipitous optical surveys can be built performing observations with 8m-class telescopes, such as the *VLT*, the *Gemini*, and with the *HST*. Such surveys would be surely incomplete in terms of color coverage, but they would give a valuable support for some identification works if the science data centers would provide the advanced data products - e.g. fully reduced and calibrated frames, image stacks, and object catalogs - needed for this kind of projects. The next planned surveys, such as VISTA [4], Pan-STARSS [5] and LSST [6], will surely provide useful data for optical identifications of faint high energy sources. Moreover, next generation world-class telescope, such as the E-ELT, the GMT and the TMT will ensure a major step forward in the next decade for optical/near-infrared follow-up.

Nonetheless, dedicated spectroscopic campaigns of relatively faint sources are being carried out successfully with the 3.58m Telescopio Nazionale Galileo (TNG), La Palma (Canary Islands - Spain). Here I will review two science cases for which the contribution from the TNG proved to be crucial, namely the optical spectroscopy of serendipitous candidate Active Galactic Nuclei (AGN) detected with the *Swift* satellite, and the optical follow-up of candidate blazars detected by the *Fermi* Gamma-Ray Space Telescope.

1.1. Optical follow-up of Swift serendipitous AGN candidates selected for X-ray variability studies

Statistical properties of X-ray variability may provide clues for understanding the underlying X-ray emission. In the optical domain, *ensemble* analyses based on variability of SDSS data sets provided a detailed description of the dependence of variability on luminosity, redshift, wavelength and time lag between observations, via a structure function (SF) analysis [7; 8]. Recently, a similar study in the X-ray domain [9], based on two different serendipitously selected samples extracted from the catalogs of XMM-Newton [10] and *Swift* [S3XGF; 11], gives statistically consistent results in the two cases, with the SF described by a power law of the time lag, with exponent $b = 0.10 \pm 0.01$ (XMM-Newton) or $b = 0.07 \pm 0.04$ (*Swift*). This would correspond to a Power Density Spectrum (PDS) with power law exponent ~ 1.2 for the case of a single-power-law PDS, which is within the range of exponents found for nearby Seyferts [12]. The two samples are complementary since XMM-Newton data probe timescales within the range from months to a few years in the rest-frame, while *Swift* typical monitoring times range from some days to a few weeks in the rest-frame. Moreover, XMM-Newton suffers for rather sparse sampling, providing only two epochs for the majority of sources, whereas *Swift* provides a sampling at intermediate time-scales (hours to a few months in the observed frame).

To build an *ensemble* SF of the AGN, it is necessary: i) to ascertain the AGN nature of the X-ray serendipitous sources, excluding possible X-ray emitting stars or galaxies, and ii) to measure the redshift in order to group all the individual flux variations of different objects in bins of rest-frame time lag. Analysis of *Swift* GRB fields, observed between January 2005 and June 2007, overlapping the area covered by the Data Release 7 of the Sloan Digital Sky Survey [SDSS-DR7 13] allowed us to identify only 27 sampled-well-enough (i.e. at least, 100 photons in the lightcurve) serendipitous X-ray sources with quasi-stellar-objects (QSOs) of known redshift to built the SF shown in Fig. 6 of [9], providing a curve of growth of the X-ray variability with respect to rest-frame time lag $\tau \leq 10$ days.

To extend the *Swift* SF to have a better sampling at longer timescales, I lead a dedicated spectroscopy campaign to observe a group of serendipitous S3XGF sources, selected to have long exposure, a good sampling of the X-ray lightcurve and an optical counterpart in the SDSS-DR7 brighter than $g' = 21.5$. Spectroscopic observations were carried out on July 15/16 2010 with the DOLORES spectrograph at the TNG. DOLORES Grism LR-B and long-slit mode, with 1.5 arcsec slit, were chosen to have the spectral resolution needed to distiguish broad and narrow emission lines. I give some information on data reduction in Appendix A.

15 objects were observed: among them, the AGN nature of 11 sources ($\sim 80\%$) was confimed, and their redshifts were measured. These new identifications will enable the extension of the previously published X-ray SF up to rest-frame time lag of $\tau \sim 40$ days [14].

1.2. Optical identification of candidate blazars discovered by the Fermi Gamma-Ray Space Telescope

The *Fermi* Gamma-Ray Space Telescope was successfully launched in June 2008 and is now in full operational mode providing an unprecedented view of the high-energy γ-ray sky. One of *Fermi*'s primary goals is to investigate the physics of relativistic jets in AGN, to which end it is already making good progress towards an order-of-magnitude increase in the number of known γ-ray bright blazars. Blazars are rare extragalactic objects as they only make a fraction of radio loud AGN, which, in turn are much less abundant (10%) than the bulk of the AGN population. Typical AGN are radio-quiet and are found in large numbers at optical and X-ray frequencies. However, blazars strong emission at all wavelengths makes them the dominant type of extragalactic sources in those energy windows where the accretion onto a supermassive black hole, or other thermal mechanisms, do not produce much radiation. For instance, in the microwave band, [15] showed that blazars are the largest population of extragalactic objects. The same is true in the γ-ray band [16; 17], and at TeV energies where BL Lacs are the most frequent type of sources found in the high Galactic latitude sky [e.g. 17; 18, see e.g. the Web-based TeVCat 1 catalog for an updated list of TeV sources].

Blazars have been known and studied in different energy windows for over 40 years, however, many questions still remain open about their physics and demography, e.g., how are jets made and how are they accelerated? What is the jet matter content? How are the relativistic particles accelerated and why is the maximum acceleration so much higher in BL Lacs? What is the maximum acceleration that can be achieved? What are the mechanisms producing blazar variability and what is the blazar duty-cycle? What is the cosmological evolution of the different types of blazars, namely BL Lacs and Flat Spectrum Radio Quasars (FSRQs)?

Fermi , with its large sensitivity and unique capability of observing the entire sky every day, started a new era in γ-ray astronomy. Its great and unprecedented potential will lead to the detection of very large and statistically well defined samples, as well as to the monitoring of hundreds of blazars on different time scales and on a wide γ-ray energy band. These new γ-ray data sets, together with the others at different energy bands will be used to address many, if not all, of the questions listed above.

All types of blazars emit radiation across the entire electromagnetic spectrum through the widely recognized Synchrotron Self Compton (SSC) mechanism which is responsible for the well known double hump Spectral Energy Distribution (SED). However FSRQs and BL Lacs differ at least in i) the optical band, where the former show the broad lined spectrum typical of all QSOs whereas BL Lacs are featureless, and in ii) the energetics of the emitting particles with BL Lacs reaching in some cases much higher energies shifting to higher frequencies the peak of their synchrotron and inverse Compton power. Blazars can therefore be distinguished by the peak of their synchrotron emission, that is the main reason why one of the most effective ways of studying their properties is through the use of multi-frequency data.

The first year of sky-survey operation with the *Fermi* Large Area Telescope (LAT) revealed 1079 bright sources at high galactic latitude ($|b| > 10$) indicating high-confidence associations of 700 of these sources with known AGN. About 20% of these are still unidentified [1LAC; 19].

In this context, a long observational campaign, led by S. Piranomonte, started at TNG in 2009, and I joined the team in 2010. The main goal was to complete the identification of the "northern" unidentified sources from 1LAC and to include them in the 2nd year *Fermi* LAT AGN catalogue [2LAC; 20].

Optical spectroscopy was then obtained for a wide sample of unidentified sources in the "northern" sky that show the typical hallmarks of blazars, such as radio loudness, flat radio spectrum etc, and are statistically associated (probability larger than 90%) with one of the still unclassified 1LAC γ-ray sources. In particular, 54 unclassified extragalactic LAT sources were observed with DOLORES at TNG between 2009-2010: 16 were already updated in 1LAC, and

21 new redshifts were measured (7 added in 1LAC, 14 updated recently in 2LAC). Other 10 unclassified LAT sources were selected to be observed by the end of march 2011. Different grisms and slits were chosen to achieve the best results. Data reduction is detailed in Appendix A. The campaign confirmed the blazar nature of all the over 60 candidates: among them, 85 % are BL Lacs (20% with redshift), the rest are FSRQs (all with redshift) [21]. These spectroscopic identifications were crucial to identify all of the still unidentified high latitude radio loud sources in 1LAC and to insert them in the 2LAC, together with the results coming from similar campaigns carried at the Hobby-Eberley Telescope (HET), the 2.7m at McDonald Observatory, the NTT at La Silla, the 200″ Hale Telescope at Mt. Palomar, the VLT, and the W. M. Keck Observatory [22–24].

The improved knowledge of the *Fermi* -LAT γ-ray sources will shortly lead to the construction of very large, statistically complete and absorption-unbiased samples of blazars enabling for the first time detailed studies of this puzzling class of sources.

Acknowledgments

I would like to thank all the people involved in these projects, especially F. Vagnetti, V. Zitelli, D. Trevese, S. Piranomonte, V. D'Elia e B. Sbarufatti.

This work is based mainly on optical spectroscopy performed at the Telescopio Nazionale Galileo (TNG), La Palma, Canary Islands (*Swift*: proposal AOT21/10A TAC9; *Fermi*: proposals AOT20/09B TAC47, AOT21/10A TAC38 and AOT23/2011A TAC40). I thank all the staff working at TNG, especially W. Boschin.

Part of this work is based on archival data, software or on-line services provided by the ASI Science Data Center (ASDC). Additional data have been obtained from the NASA/IPAC Extragalactic Database (NED) and from the Sloan Digital Sky Survey SDSS-DR7 SkyServer.

IRAF is distributed by the National Optical Astronomy Observatories, which are operated by the Association of Universities for Research in Astronomy, Inc., under cooperative agreement with the National Science Foundation.

Appendix A.

All the spectra acquired with DOLORES at TNG were processed via standard data reduction using different IRAF packages [25], to obtain 1-dimensional calibrated extracted spectra: raw data were bias-subtracted and flat-fielded using tasks in the IRAF package `noao.imred.ccdred`, and the spectra were extracted, wavelenght- and flux-calibrated using programs in the package `noao.twodspec`. Reference arcs were used to perform wavelenght-calibration. Proper flux-calibration was obtained using spectrophotometric standard stars chosen from [26]. Spectra were cleaned of cosmic rays. Emission lines were visually identified and redshifts were estimated from these lines using the task `rvidlines` within the IRAF package `noao.rv`.

References

[1] Bignami G F, Caraveo P A, Mereghetti S, Paul J, Cordier B, Goldwurm A, Mandrou P, Roques J P and Vedrenne G 1991 *The Messenger* **66** 10–13

[2] Mignani R P 2009 *ArXiv e-prints* (*Preprint* `0902.0634`)

[3] York D G *et al* 2000 *AJ* **120** 1579

[4] Worswick S P, Atad-Ettedgui E, Casali M M and Henry D M 2000 *Society of Photo-Optical Instrumentation Engineers (SPIE) Conference Series* (*Society of Photo-Optical Instrumentation Engineers (SPIE) Conference Series* vol 4003) ed P Dierickx pp 373–380

[5] Kaiser N, Burgett W, Chambers K, Denneau L, Heasley J, Jedicke R, Magnier E, Morgan J, Onaka P and Tonry J 2010 *Society of Photo-Optical Instrumentation Engineers (SPIE)*

Conference Series (*Society of Photo-Optical Instrumentation Engineers (SPIE) Conference Series* vol 7733)

[6] Ivezic Z *et al* 2009 *American Astronomical Society Meeting Abstracts* (*American Astronomical Society Meeting Abstracts* vol 213) pp 460.03–+

[7] Vanden Berk D E et al 2004 *ApJ* **601** 692–714

[8] de Vries W H, Becker R H, White R L and Loomis C 2005 *Astron. J.* **129** 615–629

[9] Vagnetti F, Turriziani S and Trevese D 2011 *A&A* **536** A84 (*Preprint* 1110.4768)

[10] Watson M G *et al* 2009 *A&A* **493** 339–373 (*Preprint* 0807.1067)

[11] Puccetti S et al 2011 *A&A* **528** A122+ (*Preprint* 1102.4270)

[12] Lawrence A and Papadakis I 1993 *Astrophys. J. Lett.* **414** L85–L88

[13] Abazajian *et al* 2009 *ApJS* **182** 543–558 (*Preprint* 0812.0649)

[14] Turriziani S, Zitelli V, Vagnetti F, Trevese D *et al* 2012 An ensemble study of the x-ray variability of agn: optical spectroscopy, in preparation, to be published on A&A

[15] Giommi P and Colafrancesco S 2004 *A&A* **414** 7

[16] Hartman R C *et al* 1999 *APJS* **123** 79

[17] Abdo A A *et al* 2009 *ApJ* **700** 597–622 (*Preprint* 0902.1559)

[18] Giommi P, Colafrancesco S, Cavazzuti E, Perri M and Pittori C 2006 *A&A* **445** 843

[19] Abdo A A *et al* 2010 *ApJ* **715** 429–457 (*Preprint* 1002.0150)

[20] Ackermann M *et al* 2011 *ApJ* **743** 171 (*Preprint* 1108.1420)

[21] Piranomonte S, Turriziani S, D'Elia V and Sbarufatti B 2012 Optical spectroscopy of "northern sky" fermi lat blazars in preparation, to be published in A&A

[22] Shaw M S, Romani R W, Healey S E, Cotter G, Michelson P F and Readhead A C S 2009 *ApJ* **704** 477–484 (*Preprint* 0908.2996)

[23] Shaw M S *et al* 2010 *American Astronomical Society Meeting Abstracts #215* (*Bulletin of the American Astronomical Society* vol 42) p 434.12

[24] Shaw M S, Romani R W, Cotter G, Healey S E, Michelson P F, Readhead A C S, Richards J L, Max-Moerbeck W, King O G and Potter W J 2012 *ApJ* **748** 49 (*Preprint* 1201.0999)

[25] Tody D 1986 *Instrumentation in astronomy VI; Proceedings of the Meeting, Tucson, AZ, Mar. 4-8, 1986. Part 2 (A87-36376 15-35). Bellingham, WA, Society of Photo-Optical Instrumentation Engineers, 1986, p. 733.* ed Crawford D L pp 733–+

[26] Oke J B 1990 *AJ* **99** 1621–1631

Galaxy clusters mass measurements: news in X-ray and Weak lensing data analysis

Iacopo Bartalucci[1], **Ilaria Formicola**[1], **Rossella Martino**[2]

[1] Department of Physics, Università di Roma Tor Vergata, 00133 Rome, Italy
[2] Laboratoire IRFU/Service dAstrophysique - CEA - CNRS - Bat. 709, CEA-Saclay, F-91191 Gif-sur-Yvette Cedex, France

E-mail: bartalucci@roma2.infn.it, formicola@roma2.infn.it, martino@roma2.infn.it

Abstract.
We describe the recent improvements made on the techniques for the estimates of the total mass of galaxy clusters using the X-ray data, obtained by the Chandra and XMM satellites, and the weak lensing observations, performed by Subaru telescope. Furthermore, we show the current state of X-ray and weak lensing mass comparison obtained by simulations and observational data. In particular, we use a subsample of 21 galaxy clusters of the LOCUSS High-L_x sample, for which e find that X-ray masses overcome weak lensing mass by about 10%.

1. Introduction

Galaxy clusters are the largest and most massive collapsed objects in the Universe. They are directly linked to the primordial density perturbations, therefore they are powerful tools to test the cosmological theories and to constrain cosmological parameters ([1], [2], [3], [4], [5]). Structures formation on large scale and their time evolution can be investigated using N-body simulations ([6], [7], [8]). Results obtained from N-body simulations need to be compared with observations in order to study hidden biases, quantifying sources of uncertainty in the measure with a good control on systematic errors. This issue can be investigated using both observational data and numerical simulations ([9], [10], [11]). Furthermore, it is very important to perform a multi-wavelength study, comparing results obtained from two different techniques based respectively on X-ray and weak lensing observations ([12], [13], [14]).
The X-ray emission produced by the hot Intracluster Medium (ICM) within galaxy clusters allows to estimate cluster mass profiles analyzing the gas density and temperature structures, assuming the hot ICM to be in hydrostatic equilibrium within the cluster gravitational potential well.
The weak lensing analysis measures weak distortions in the shape of faint distant galaxies, induced by the cluster potential, to derive the mass. Weak gravitational lensing provides a unique tool to study the dark matter content in galaxy clusters since it probes the total matter density with no distinction between ordinary barionic matter and dark matter and without any assumption on the shape and/or on the dynamical state of the cluster. However, halo triaxiality could lead to an overestimation of mass and concentration derived from the NFW halo profile. In fact in this case a measure of the three-dimensional structure of the halo is obtained from a 2-D information, since lensing is sensitive only to the projected mass ([15]). On the other

hand, with respect to the weak lensing analysis, the X-ray method has the advantage to be less sensitive to projection effects due to additional mass concentration along the line of sight. All these reasons make complementary the two techniques. With the aim to derive an accurate mass measure using both techniques we describe the main features of the recent improvements made for each of them. In particular, we develop a new analytical background model for Chandra data, allowing to study in detail galaxy clusters up to their outskirts, and test selection criteria for background lensed sources, in order to avoid dilution in the lensing signal, using lensing simulation and applying these criteria to Subaru data. Furthermore, we show the current state of X-ray and weak lensing mass comparison obtained from simulations and observational data.

2. Weak Lensing Analysis

An accurate cluster mass measure by weak lensing requires a careful selection of background galaxies. Therefore it's necessary to avoid, in the selected sample, cluster and foreground unlensed sources which dilute the lensing signal. Dilution increases as the cluster-centric distance decreases since the number density of cluster galaxies, which contaminate the background galaxies sample, follows the density profile of the cluster. Without any redshift information, it's possible to select background sources using photometry or photometric redshifts depending on the number of photometric bands available. Starting from magnitudes in three bands, cluster, background and foreground galaxies can be identified as overdensities in two-colors diagrams as suggested by [16].

In this section we describe our criteria for selection of galaxies in two-colors diagrams, applied to simulations of lensing observations and to a real cluster: Abell 2219. Wide field simulated images of the g_1 cluster ([17]) in B, V, R, I, z bands using the SUBARU Suprime-Cam allow us to derive galaxy colors and photometric redshifts. Our selection criteria, based on galaxy colors, use the COSMOS catalog with photometry and photometric redshifts ([18]) to derive where galaxies with photometric redshifts lower or higher than the cluster redshift lie in color-color diagrams, obtained plotting the $V-R$ vs $B-R$ and the $R-I$ vs $B-V$ colors. We apply the same color cuts which identify each population of galaxies in the COSMOS two-colors diagrams to sources belonging to the images of the g_1 cluster, deriving in this way the background and the foreground galaxy populations. In order to test our selection we derive galaxy radial density profiles and shear profiles of both tangential and radial components for each identified population. A good selection, indeed, implies tangential shear profiles for cluster and foreground galaxies consistent with zero and almost flat galaxy radial density profiles for background and foreground galaxies, since these galaxies are uniformly distributed across the field of view. The shear signal is obtained averaging on the observed ellipticities, corrected for the observational distortions introduced by seeing and PSF anisotropy. This correction is done following the approach suggested in the work of [19], (KSB method), and performed as described in [20] and [21].

Real redshift and ellipticities, available for each source in the images of the g_1 cluster, allow us to quantify the contamination due to unlensed sources both in the case of a selection based on colors and in the case of a selection based on photometric redshifts, derived using the photometry in five bands.

We find that less than the 10% of foreground galaxies contaminate the background color-selected sample and the background sample selected with photometric redshifts. In addition, as described in [22], the KSB approach underestimates measured ellipticities: for these simulations measured ellipticities are underestimated of about 20% with respect to true ellipticities.

On the basis of results obtained from simulations we apply the described selection methods to a real cluster, Abell 2219, in order to derive an accurate mass estimation by weak lensing. Abell 2219 is a massive X-ray luminous cluster at $z = 0.22$. Cluster observations were performed with the SUPRIME cam at the SUBARU telescope in B, V, R_c bands and with the Large Binocular

Camera at the LBT in I band. Subaru images were reduced with VST-Tube imaging pipeline ([23]) using ASTROMC ([24]) to remove distortions introduced by optics. The observed shear of background galaxies, selected as described above, is used to derive an estimate of the cluster mass following the aperture densitometry technique ([25]) and fitting the expected shear profile described by SIS and NFW models to data ([26]). More details on described methods and obtained results will be found in [27].

3. X-Ray Technique

In this section we explain the main features for the mass estimates using the X-ray data, focusing on the news in our technique. We use data obtained from the X-ray Chandra and XMM-Newton satellites, with the ACIS-I (Advanced CCD Imaging Spectrometer)[1] and EPIC (European Photon Imaging Camera) [2] cameras respectively. The imaging and the spectral analysis for Chandra ACIS-I and XMM-Newton EPIC observations are performed following the procedure described in [28] and [14], where all event-lists are merged in a single energy position photon cube. The brightness and temperature profiles are extracted in concentric annuli centered on the centroid of the cluster and then are used to derive the density and 3-D temperature profiles following the "forward" method described e.g. in [10]. In particular we assume a modified β-model profile for the density distribution and a power-law for the temperature profile as described in [29] and [14]. These profiles are then projected along the line of sight, in concentric shells. For XMM-Newton data, the projection of density includes a convolution by the instrument PSF. The density and 3-D temperature profiles are used to estimate the total gravitating mass assuming the hydrostatic equilibrium condition: $-G\mu m_p n_{gas} M_{tot}(< r)/r^2 = dP/dr = d(n_{gas} \times T)/dr$, where G is the gravitational constant, μ= 0.59 the mean molecular weight in a.m.u., m_p the proton mass, k the Boltzmann constant, and n_{gas} and T are the 3D gas density and temperature profiles respectively.

In this technique to measure projected temperature we need to separate adequately the *background* components spectra from the thermal emission spectra of the ICM. When we refer to background we mean the sum of two components: the physical background, produced by X-ray sources, and an instrumental background, produced by the high energy particles which hit detectors and cause instruments to glow. The first component is well known and analytically modeled ([30], [31]). The second component is modeled only for XMM-Newton observations ([31], [32]). In our work we focus on the analysis of Chandra ACIS-I particle background. The Chandra standard analysis tools provide background datasets that are subsequently subtracted from the observation of interest causing the eventual mixing of systematic and measure errors. For instance we know that instrumental background has a spatial variation gradient ([33]) which introduces systematic errors in the normalization of the background components. In the central regions the signal level is typically $2-3$ order of magnitude greater than the background signal so the mixing effect is negligible, while it becomes important in the outskirts where the ICM emission is faint and the signal level is of the same level of background. Studying galaxy cluster outskirts is crucial to fully understand evolution and formation of this structures. For the first time we produce an analytical model for Chandra ACIS-I instrument which allows us to predict particle background, minimizing systematic error, with a precision up to 2% in the continuum and 5% in the lines.

[1] for more details see http://cda.harvard.edu/chaser/

[2] for more details see http://xmm.esac.esa.int/xsa/

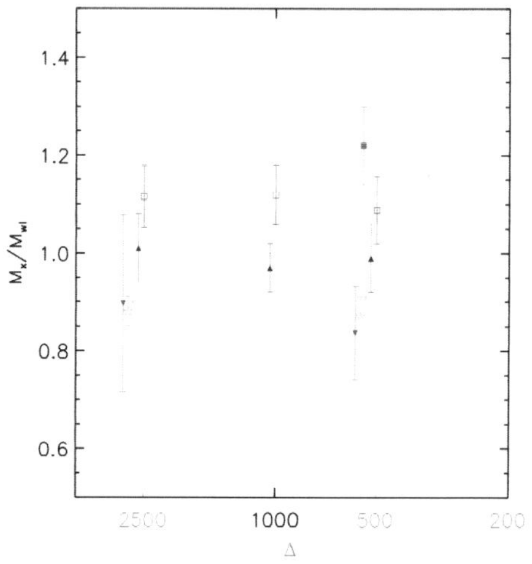

Figure 1. *The X-ray hydrostatic to weak lensing mass ratios as a function of density contrast for our subsample (black squares) are plotted. The values obtained by [12] sample (blue triangles), by [10] sample with forward method (yellow stars) and backward method (orange stars) and values obtained by [11] (magenta upside triangles) are also plotted.*

4. Results and Comparisons

We apply our X-ray technique to derive the total hydrostatic mass for the 51 clusters of galaxies of the LOCUSS High-L_X volume limited sample. We analyze all the available Chandra and XMM-Newton observations for this sample. As it will be shown in details in [14], using a subsample of 22 clusters observed with both instruments, the total cluster masses are totally consistent with an accuracy of the 5%. Therefore, for a subsample of 21 clusters we compare our X-ray hydrostatical masses with the most recent observational results of weak lensing mass estimates, presented in [34], at $r = r_\Delta$, where $\Delta = 2500, 1000, 500$ is defined as the ratio of cluster matter density over the critical density to have a flat Universe. The average X-ray to weak lensing mass ratios are calculated taking into account the errors in the X-ray and weak lensing mass estimates, following the method described in [35]. We find that the mass ratios M_X/M_{WL} are equal to 1.12 ± 0.06, 1.12 ± 0.06 and 1.09 ± 0.07 at $\Delta = 2500$, 1000 and 500, respectively. We find that X-ray masses are higher of a 10% than weak lensing masses at all overdensities. In Fig 1 we plot our X-ray hydrostatic to weak lensing mass ratios as a function of density contrast as black squares. Therefore, we compare our results with previous ones and numerical simulations. In Fig 1 previous observational results obtained by [12] are plotted as blue triangles: in this work, the X-ray and weak lensing masses are in agreement, with a mass ratio equal to the unity for all overdensities. We also overplot the recent results obtained by *Pratt et al. 2012* at r_{500} with green squares. In this last work, they find that the hydrostatic X-ray masses are on average $22 \pm 8\%$ larger than the corresponding weak lensing masses. They show how this results seem to be due to the offset of the centre of the cluster used in the analysis, equal to X-ray peak and BCG position for x-ray and weak lensing techniques respectively. We also plot the previous numerical simulations results obtained by [10] sample with forward method (yellow stars) and backward method (orange stars) and values obtained by [11] (magenta upside triangles). Simulations find that X-ray masses are low-biased compared to lensing masses by 10-20% at all overdensities. These results have an opposite trend with respect to our measurements, for which X-ray hydrostatical masses are, on average, higher than weak lensing masses of 10%.

Therefore, the current comprehension of the differences between X-ray and weak lensing masses is yet puzzling. This is mostly due to the non-availability of large samples with both high-quality X-ray and weak lensing mass estimates. This is one of the main aims of the Locuss project, and new results should be obtained in few years.

5. References

[1] Press and Schechter 1974 *Astrophys. J.* **187** 425-438
[2] S. White, F. Navarro , E. Evrard and C. Frenk 1993 *Nature* **366** 429-433
[3] V. R. Eke, and S. Cole, and C. S Frenk, and J. Patrick Henry, 1998 *Mon. Not. R. Astron. Soc* **298** 1145-1158
[4] S. W. Allen, and R. W. Schmidt, and A. C. Fabian, 2002 *Mon. Not. R. Astron. Soc* **334** L11-L15
[5] A. Vikhlinin, and A. V Kravtsov, and R. A. Burenin, and H. Ebeling, and W. R. Forman, and A. Hornstrup, and C. Jones, and S. S. Murray, and D. Nagai, and H. Quintana, and A. Voevodkin, 2009 *Astrophys. J.* **692** 1060-1074
[6] R. K. Sheth, and G. Tormen, 1999 *Mon. Not. R. Astron. Soc* **308** 119-126
[7] A. Jenkins, and C. S. Frenk, and S. D. M. White, and J. M. Colberg, and S. Cole, and A. E. Evrard, and H. M. P. Couchman, and N. Yoshida, 2001 *Mon. Not. R. Astron. Soc* **321** 372-384
[8] C. Giocoli, and G. Tormen, and F. C. van den Bosch, 2008 *Mon. Not. R. Astron. Soc* **386** 2135-2144
[9] E. Rasia, and S. Ettori, and L. Moscardini, and P. Mazzotta, and S. Borgani, and K. Dolag, and G. Tormen, and L. M. Cheng, and A. Diaferio, 2006 *Mon. Not. R. Astron. Soc* **369** 2013-2024
[10] M. Meneghetti, and E. Rasia, and J. Merten, and F. Bellagamba, and S. Ettori, and P. Mazzotta, and K. Dolag, and S. Marri, 2010 *Astron. & Astrophys* **514** A93
[11] D. Nagai, and A. Vikhlinin, and A. V. Kravtsov, 2007 *Astrophys. J.* **655** 98-108
[12] Y.-Y. Zhang, and N. Okabe, and A. Finoguenov, and G. P.Smith, and R. Piffaretti, and R. Valdarnini, and A. Babul, and A. E. Evrard, and P. Mazzotta, and A. J. R. Sanderson, and D. P. Marrone, 2010 *Astrophys. J.* **711** 1033-1043
[13] Planck Collaboration 2012 *ArXiv e-prints* 2743P
[14] R. Martino et al. in preparation
[15] V. L. Corless, and L. J. King, 2007 *MNRAS* **380** 149-161
[16] E. Medezinski, and T. Broadhurst, and K. Umetsu, and M. Oguri, and Y. Rephaeli, and N.Benítez, 2010 *MNRAS* **405** 257-273
[17] M. Meneghetti, and R. Argazzi, and F. Pace, and L. Moscardini, and K. Dolag, and M. Bartelmann, and G. Li, and M. Oguri, 2007 *Astron. & Astrophys* **461** 25-38
[18] O. Ilbert, and P. Capak, and M. Salvato, and H. Aussel, et al. 2009 *Astrophys. J.* **690** 1236-1249
[19] N. Kaiser, and G. Squires, and T. Broadhurst, 1995 *Astrophys. J.* **449** 460
[20] Z. Huang, and M. Radovich, and A. Grado, and E. Puddu, and A. Romano, and L. Limatola, and L. Fu, , 2011 *Astron. & Astrophys* **529** A93
[21] M. Radovich, and E. Puddu, and A. Romano, and A. Grado, and F. Getman, 2008 *Astron. & Astrophys* **487** 55-61
[22] T. Erben, and L. Van Waerbeke, and R. Bertin, and Y. Mellier, and P. Schneider, 2001 *Astron. & Astrophys* **366** 717-735
[23] A. Grado, and M. Capaccioli, and L. Limatola, and F. Getman, 2012 *Image Processing, Astronomical Techniques, Astrophysical Data* **19** 362
[24] M. Radovich, and M. Arnaboldi, and V. Ripepi, and Massarotti, et al. 2004 *Astron. & Astrophys* **417** 51-60
[25] D. Clowe, and G. A. Luppino, and N. Kaiser, and J. P. Henry, and I. M. Gioia, 1998 *Astrophys. J. L.* **497** L61
[26] P. Schneider, and L. King, and T. Erben, T. 2000 *Astron. & Astrophys* **353** 41-56
[27] Formicola et al. in preparation
[28] P. Mazzotta, and S. Giacintucci, 2008 *Astrophys. J.* **675** L9-L12
[29] A. Vikhlinin, and A. Kravtsov, and W. Forman, and C. Jones, and M. Markevitch, and S. S. Murray, and L. Van Speybroeck, 2006 *Astrophys. J.* **640** 691-709
[30] D. H. Lumb, and R. S. Warwick, and M. Page, and A. De Luca, 2002 *Astron. & Astrophys* **389** 93-105
[31] S. L. Snowden, and M. J. Freyberg, and K. D. Kuntz, and W. T. Sanders, 2000 *Astrophys. J.* **128** 171-212
[32] A. Leccardi, and S. Molendi, 2008 *Astron. & Astrophys* **486** 359-373
[33] Bartalucci et al. in preparation
[34] N. Okabe, and M. Takada, and K. Umetsu, and T. Futamase, and G. P. Smith, 2010 *Publications of the ASJ* **62** 811
[35] A. Mahdavi, and H. Hoekstra, and A. Babul, J. P. and Henry, 2008 *Mon. Not. R. Astron. Soc* **384** 1567-1574

Forecasting the curvaton isocurvature scenario with CMB lensing information

Larissa Santos

Dipartimento di Fisica, Università di Roma "Tor Vergata"

E-mail: `larissa.santos@roma2.infn.it`

Abstract. Some inflationary models predict the existence of isocurvature primordial fluctuations, in addition to the well known adiabatic perturbations. Such mixed models are not yet ruled out by available data sets. The talk presented at the Young Research Meeting covered the possibility of obtaining better constraints on the isocurvature contribution from future astronomical data. I use Planck satellite experimental specifications together with SDSS galaxy survey to forecast for the best parameter error estimate by means of the Fisher information matrix formalism. In particular, I consider how CMB lensing information can improve this forecast. I focus specially in the curvaton inflationary scenario to show a substantial improvement in all the considered cosmological parameters. In the case of isocurvature amplitude the improvement is above 20% around its fiducial value. In this sense, CMB lensing information will be crucial in the analysis of future data.

1. Introduction

Since the early measurement of the first acoustic peak in the Cosmic Microwave Background (CMB) angular power spectrum [10, 14], a pure isocurvature model of primordial fluctuations was ruled out [13]. In addition, recent CMB data from the WMAP satellite found no evidence for non-adiabatic primordial fluctuations [17]. However, small contributions from isocurvature primordial fluctuations cannot be excluded by the current data.

The standard inflationary scenario driven by the inflaton cannot account for isocurvature fluctuations. We consider the alternative scenario where perturbations to the curvaton light field are responsible for curvature perturbations and may also generate isocurvature fluctuations [23, 25, 22]. In this case, the isocurvature component is completely correlated or anti-correlated with the adiabatic component. Following [1], the isocurvature contribution to the purely adiabatic CMB power spectra is written in the form:

$$C_l = (1-\alpha)C_l^{ad} + \alpha C_l^{iso} + 2\beta\sqrt{\alpha(1-\alpha)}C_l^{cross}, \quad (1)$$

where the parameter α accounts for the isocurvature amplitude, while β stands for the isocurvature correlation phase, given by $\beta = \cos\theta$, $-1 \leq \cos\theta \leq 1$. In the same way, the matter power spectrum, $P(k)$, can be decomposed into purely adiabatic, purely isocurvature and their cross correlation contribution. I assume that $n_{cross} = \frac{1}{2}(n_{ad} + n_{iso})$ as stated by [1] and take the pivot value $k_0 = 0.002 Mpc^{-1}$ as used by the WMAP team.

There are already many studies in the literature constraining the isocurvature contribution using different data sets (see, for example, [21, 6, 18, 17, 24, 1, 2, 8]). I intend to use the well-known Fisher information matrix formalism to estimate whether better constraints to the isocurvature contribution can be obtained in the near future using measurements of the CMB temperature and polarization power spectra from the Planck satellite, also using CMB lensing information, as well as the large-scale matter distribution observed by the Sloan Digital Sky Survey (SDSS).

A small introduction on CMB lensing is given in Section 2. In Section 3, a brief review of the Fisher information matrix formalism for the CMB and for a galaxy survey is presented. Finally, results are shown in Section 4, followed by the discussion and conclusions in Section 5.

2. CMB lensing

As new experiments are developed, the precision in CMB measurements makes small effects distinguishable by observations. CMB lensing is one of this effects and it has important quantitative contribution that should be taken into account. It is known that the CMB photons are deflected by gravitational potentials during their travel between the last scattering surface and the observer. For CMB temperature anisotropy, this is quantitatively written as:

$$\frac{\Delta\tilde{\mathrm{T}}(\hat{\mathbf{n}})}{T} = \frac{\Delta\mathrm{T}(\hat{\mathbf{n}}')}{T} = \frac{\Delta\mathrm{T}(\hat{\mathbf{n}}+d)}{T}, \tag{2}$$

where the temperature T of the lensed CMB in a direction $\hat{\mathbf{n}}$ is equal to the unlensed CMB in a different direction $\hat{\mathbf{n}}'$. Both these directions, $\hat{\mathbf{n}}$ and $\hat{\mathbf{n}}'$, differ by the deflection angle d as it can be seen in the third equality above. To first order, the deflection angle is simply the lensing potential gradient, $d = \nabla\psi$ (for a review of CMB lensing see [19]). In the same way, the effect of lensing in CMB polarization is written in terms of the Stokes parameters $Q(\hat{\mathbf{n}})$ and $U(\hat{\mathbf{n}})$ (for a review in CMB polarization theory, see [5]) :

$$[Q + iU](\hat{\mathbf{n}}) = [Q + iU](\hat{\mathbf{n}} + d). \tag{3}$$

We can study CMB lensing properties through the lensing potential, and the temperature and polarization power spectra, as well as their cross-correlation.

The lensing signal was detected for the first time by cross-correlating WMAP data to radio galaxy counts in the NRAO VLA sky survey (NVSS) [28]. The detection of the gravitational lensing using CMB temperature maps alone and the measurement of the power spectrum of the projected gravitational potential were already done using the Atacama Cosmology Telescope and the South Pole Telescope [9, 31]. While waiting for more precise data sets, much work is being done to CMB lensing reconstruction techniques (e.g. [15, 26, 29, 4, 7]). Here, I use the CAMB software package [20] to obtain the numerical lensed and unlensed power spectra for the chosen cosmological model and then forecast how CMB lensing information can help constraining the cosmological parameters.

3. Method

In order to search how an isocurvature contribution would affect the measurements of cosmological parameters I apply the Fisher information matrix formalism to a Planck-like experiment [3], considering both temperature and polarization for the lensend and unlensed CMB spectra, and the matter power spectrum from the Sloan Digital Sky Survey (SDSS).

3.1. Information from CMB

The Fisher information matrix for the CMB temperature anisotropy and polarization is given by the approximation in [32]:

$$F_{ij} = \sum_{l}\sum_{XY}\frac{\partial C_l^X}{\partial p_i}(Cov_l^{-1})_{XY}\frac{\partial C_l^Y}{\partial p_j}, \tag{4}$$

Journal of Physics: Conference Series **383** (2012) 012012 doi:10.1088/1742-6596/383/1/012012

where C_l^X is the power in the lth multipole, X stands for TT (temperature), EE (E-mode polarization), BB (B-mode polarization) and TE (temperature and E-mode polarization cross-correlation), and p is a parameter of the fiducial model that we want to constraint. We will not include primordial B-modes in the analysis since the measurement of the primordial C_l^{BB} by Planck is expected to be noise dominated. The covariance matrix becomes therefore:

$$Cov_l = \frac{2}{(2l+1)fsky}\begin{bmatrix} \Xi_{TTTT} & \Xi_{TTEE} & \Xi_{TTTE} \\ \Xi_{EETT} & \Xi_{EEEE} & \Xi_{EETE} \\ \Xi_{TETT} & \Xi_{TEEE} & \Xi_{TETE} \end{bmatrix}, \tag{5}$$

where $fsky$ is the observed sky fraction. For the lensed case we have to perform a correction in the covariance matrix elements taking into consideration the power spectrum of the deflection angle and its cross correlation with temperature, C_l^{Td}. Note that in this case we are taking into consideration the B-mode polarization generated by the CMB gravitational lensing from the E-mode polarization. When these corrections are included, the covariance matrix becomes:

$$Cov_l = \frac{2}{(2l+1)fsky}$$
$$\begin{bmatrix} \xi_{TTTT} & \xi_{TTEE} & \xi_{TTTE} & \xi_{TTTd} & \xi_{TTdd} & 0 \\ \xi_{EETT} & \xi_{EEEE} & \xi_{EETE} & 0 & 0 & 0 \\ \xi_{TETT} & \xi_{TEEE} & \xi_{TETE} & 0 & 0 & 0 \\ \xi_{TdTT} & 0 & 0 & \xi_{TdTd} & \xi_{Tddd} & 0 \\ \xi_{ddTT} & 0 & 0 & \xi_{ddTd} & \xi_{dddd} & 0 \\ 0 & 0 & 0 & 0 & 0 & \xi_{BBBB} \end{bmatrix}. \tag{6}$$

Full expressions for the elements of both covariance matrices can be found in [11] and [27]. The zero elements in the covariance matrix are the ones related to the cross correlated B mode power spectra, since we assume $C_l^{TB} = C_l^{EB} = C_l^{dB} = 0$. I also consider C^{Ed}=0. Note that in this case the B-mode polarization generated by the CMB gravitational lensing from the E-mode polarization has being taken into consideration. In both cases, it is assumed that the experiment covers 65% of the sky.

3.2. *Information from galaxy survey*

The Fisher information matrix for the matter power spectrum obtained from galaxy surveys is given by [30]:

$$F_{ij} = \int_{k_{min}}^{k_{max}} \frac{\partial \ln P(k)}{\partial p_i}\frac{\partial \ln P(k)}{\partial p_j} V_{eff}\frac{k^2 dk}{(2\pi)^2}, \tag{7}$$

$$V_{eff}(k) = \int \left[\frac{\bar{n}(r)P_g(k)}{1+\bar{n}(r)P_g(k)}\right]^2 d^3r. \tag{8}$$

We know that $P_g(k) = b^2P(k)$ and using the specifications of the SDSS experiment for the Bright Red Galaxy (BRG) sample, called Luminous Red Galaxies (LRG) in more recent papers, we assume a linear and scalar independent bias $b = 2$ [30, 16]. It is assumed that the expected number density of galaxies, $\bar{n}(r)$, is independent of r, $n = 10^5/V_s$, in a volume-limited sample to a depth of 10^3Mpc. The survey has an angle of π steradians; therefore the survey volume becomes, $V_s = 10^9\pi/3$ [30, 11, 12].

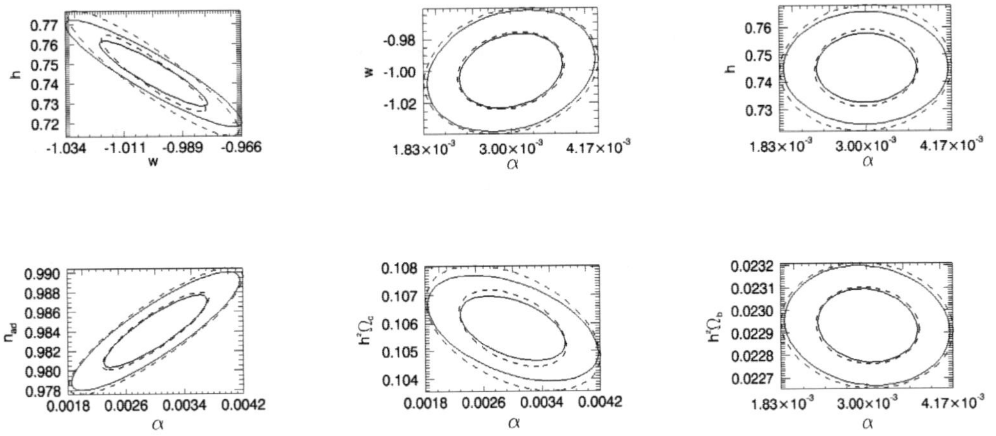

Figure 1. Fisher contours for the ΛCDM model plus a contribution of initial isocurvature fluctuations with fiducial amplitude of $\alpha = 0.003$ and correlation phase of $\beta = -1$ for a curvaton scenario. The red and blue contours represent 95.4% and 68% C.L. respectively for the unlesed CMB + SDSS (dashed lines) and for the lensed CMB + SDSS (solid lines).

4. Results

For the purpose of this analysis, I consider as free cosmological parameters the adimensional value of the Hubble constant, h, the density of baryons and cold dark matter, $\Omega_b h^2$ and $\Omega_c h^2$, the spectral index of scalar adiabatic perturbations, n_{ad}, the aforementioned isocurvature parameter, α , and the equation of state of dark energy w, assumed as a constant. I first performed the forecast for Planck alone, with and without considering CMB lensing. Then, I introduced the forecast for SDSS, combining the results as:

$$F_{ij}^{Total} = F_{ij}^{Planck} + F_{ij}^{SDSS}. \tag{9}$$

The fiducial parameters' values are set to be, $h = 0.745$, $\Omega_b = 0.02293h^2$, $\Omega_c = 0.1058h^2$, $n_{ad} = 0.984$ ($n_{iso} = 0.984$ fixed), $\alpha = 0.003$, $\beta = -1$ and $w = -1$. The best constraint is reached considering the CMB lensing effect in the analysis (see, Table 1 and Figure 1), improving Planck limit to $\alpha < 0.0054$ (95% CL). For Planck (including CMB lensing) + SDSS, $\alpha < 0.0039$ (95% CL) against $\alpha < 0.0037$ (95% CL) for WMAP + BAO + SN. Adding therefore other cosmological probes, such as BAO and SN to Planck with lensing information + SDSS, the error bars in the isocurvature amplitude can become even smaller, allowing to a very limited isocurvature contribution in the primordial fluctuations.

5. Discussion and conclusions

In this presentation I considered a possible contribution of isocurvature initial perturbations in the pure adiabatic fluctuations scenario from the well tested ΛCDM model. Using the Fisher formalism I obtained the best constraints possible for the isocurvature parameters using CMB and galaxy distribution information. The main goal of this work has been to quantify how CMB lensing information can provide better constraints in the cosmological parameters, specially in the ones related to the isocurvature contribution.

In the tested inflationary scenario, the CMB lensing information improves the constraints of all chosen parameters, including the ones related to the isocurvature modes. If we consider

Table 1. Marginalized errors for ΛCDM model plus the contribution of initial isocurvature fluctuation driven by the curvaton field.

Parameter	CMB alone Planck T + P	CMB alone Planck T+ P+ lens	P(k) alone SDSS	CMB + P(k) Planck + SDSS T + P	CMB + P(k) Planck +SDSS T + P + lens
h	0.055	0.0345	0.31	0.0092	0.0083
$h^2\Omega_b$	0.00012	0.00011	0.027	0.00011	0.00011
$h^2\Omega_c$	0.0010	0.00088	0.092	0.00091	0.00077
n_ad	0.0027	0.0025	0.24	0.0026	0.0024
α	0.0019	0.0012	0.17	0.00047	0.00045
w	0.14	0.088	7.47	0.016	0.015

Percentage of the parameters' fiducial values for each error above

Parameter	CMB alone Planck T + P	CMB alone Planck T+ P+ lens	P(k) alone SDSS	CMB + P(k) Planck + SDSS T + P	CMB + P(k) Planck +SDSS T + P + lens
h	7.38%	4.63%	41.61%	1.23%	1.11%
$h^2\Omega_b$	0.52%	0.48%	117.75%	0.48%	0.48%
$h^2\Omega_c$	0.94%	0.83%	87.90%	0.86%	0.72%
n_ad	0.27%	0.25%	24%	0.26%	0.24%
α	63.33%	40%	not constrained	15.67%	15%
w	14%	8.8%	not constrained	1.6%	1.5%

Planck information alone the improvement on α reaches almost 25% around its fiducial value (see the lower part of Table 1).

When the combined Planck + SDSS forecast is done, the improvement with the use of lensing information is not so significant. This is due to the poor ability of SDSS to constrain the parameters compared to Planck, especially when CMB lensing information is included. For a CMB experiment alone, or combined with any other precise experiments on galaxies distribution, lensing is an important extra information in the attempt to know how well observations can constrain the presence of isocurvature contribution to the primordial fluctuations.

References

[1] Bean, R., et al., 2006, Physical Review D, 74, 063503
[2] Beltrán, M., et al., 2005, Physical Review D, 72, 103515
[3] http://www.rssd.esa.int/SA/PLANCK/docs/Bluebook-ESA-SCI(2005)1_V2.pdf
[4] Bucher, M., et al., 2010, MNRAS, 407, 2193-2206
[5] Cabella, P. & Kamionkowski, M., 2004, ArXiv Astrophysics e-prints, astro-ph/0403392
[6] Carbone, C., et al., 2011, Cosmology Astropart. Phys., 9, 28
[7] Carvalho, C. S., & Tenero, I., 2011, Physical Review D, 84, 063001
[8] Crotty, P., et al., 2003, Physical Review Letters, 91, 171301
[9] Das, S., et al., Phys. Rev. Lett., 107, 021301 (2011).
[10] de Bernardis, P., et al., 2000, Nature, 404, 955-959
[11] Eisenstein, D., et al., 1999, Physical Review D, 518,2-23
[12] Eisenstein, D., et al., AJ, 122, 2267-2280 (2001).
[13] Enqvist, K., et al., 2000, Physical Review D, 62, 103003
[14] Hanany, S., et al., 2000, Astrophysical Journal Letters, 545, L5-L9
[15] Hu, W., 2001, Astrophysical Journal, 557, L79
[16] Hütsi, G., 2006, Astronomy and Astrophysics, 449, 891-902
[17] Komatsu, E., et al., 2011, Astrophysical Journal Supplement, 192,18
[18] Larson, D., et al., 2011, Astrophysical Journal Supplement, 192, 16

[19] Lewis, A. & Challinor, A., 2006, Phys. Rep., 429, 1-65
[20] Lewis, A. et al., 2000, Astrophysical Journal, 538, 473-476
[21] Li, H. et al., 2011, Physical Review D, 83, 123517
[22] Lyth, D. et al., 2003, Physical Review D, 67, 023503
[23] Lyth, D. & Wands, D., 2002, Physics Letters B, 524, 5-14
[24] Mangilli, A. et al., 2010, Cosmology Astropart. Phys., 10, 9
[25] Moroi, T. & Takahashi, T., 2002, Physical Review D, 66, 063501
[26] Okamoto, T. & Hu, W., 2003, Physical Review D, 67, 083002
[27] Perotto, L., et al., 2006, Cosmology Astropart. Phys., 10,13
[28] Smith, K., M., et al., 2007, Physical Review D, 76,043510
[29] Smith, K., M., et al., 2010, ArXiv e-prints, astro-ph/1010.0048
[30] Tegmark, M., 1997, Physical Review Letter, 79, 3806
[31] van Engelen , A., et al., ArXiv: astro-ph/1202.0546.
[32] Zaldarriaga, M. & Seljak, U., 1997, Physical Review D, 55, 1830-1840

Journal of Physics: Conference Series **383** (2012) 012013 doi:10.1088/1742-6596/383/1/012013

X-ray polarimetry towards high energy and solar science

S. Fabiani[1,2], R. Bellazzini[3], F. Berrilli[1], R. Campana[2], E. Costa[2], E. Del Monte[2], F. Muleri[2], A. Rubini[2], P. Soffitta[2]

[1]Università degli studi di Roma "Tor Vergata", via della Ricerca Scientifica 1, 000133 Rome, Italy
[2]INAF-IAPS, via del Fosso del Cavaliere 100, 000133 Rome, Italy
[3]INFN, Largo B. Pontecorvo 3, I-56127 Pisa, Italy

E-mail: `sergio.fabiani@iaps.inaf.it`

Abstract. In the history of X-ray astronomy the only polarimetric measurement obtained with high significance dates back to the late '70s, when the Crab Pulsar Wind Nebula was observed. X-ray polarimetry remains a widely unexplored scientific field so far. The new 2–10 keV polarimetry era will be opened by GEMS satellite in the next future, while the extension to higher energies is still a challenging goal. The photoelectric polarimeter Gas Pixel Detector (GPD) could be employed with an Ar based gas mixture to measure the solar flares X-ray polarization up to about 35 keV, while coupling it with a Compton scattering polarimeter it would be possible to extend the energy range of measurements to higher energies.

1. Introduction

Polarimetry is an additional tool in the X-ray astronomy besides imaging, spectroscopy and timing which can be employed to study celestial sources. Polarimetric information can be useful to disentangle between source models which otherwise would not be distinguished depending on the other observational techniques. Historically the only polarimetric measurement obtained with high significance dates back to the late '70s [1, 2]. It gave the confirmation that the Crab Nebula X-ray emission was due to synchrotron radiation ranging from radio wave to X-rays. In the next future X-ray polarimetry will be performed by GEMS [3, 4, 5] in the 2-10 keV energy band. This satellite mission will be equipped with two gas polarimeters based on the Time Projection Chamber (TPC) technology exploiting the dependence of the photoelectric cross section on the photons polarization. Recently many studies have been proposed to extend X-ray polarimetry towards higher energies. Photoelectric polarimeters can employ high Z gas mixtures [6, 7], while Compton scattering polarimeter are needed to measure polarization up to hundreds of keV [8, 9, 6, 10, 11, 12]. The polarimetry extension up to the hard X-ray energy band allows to study the solar flares physics [7, 13]. As for other sources the measurements of X-ray polarization from solar flares did not give exhaustive results so far [14, 15, 16]. In this work we will show the possibility to employ the photoelectric polarimeter Gas Pixel Detector (GPD) to measure X-ray polarization from solar flares up to 35 keV exploiting also its imaging capabilities. To extend the energy band up to higher energy it is possible to couple the GPD with a Compton polarimeter.

2. Polarimetry Basics

A polarimeter can be thought as an analyser of angular directions coupled with a detector. The analyser allows to distinguish the angular directions of polarization and the detector allows to detect the signal for each angular direction. Therefore the instrument response will depend on the radiation preferential angular direction of polarization. If the detected radiation is unpolarized there are no preferential angular directions therefore the polarimeter response will be flat, being the same for each analysed angular direction (see figure 1a). If the detected radiation is polarized, the instrument will have a modulated response (see figure 1b). Both the photoelectric effect and

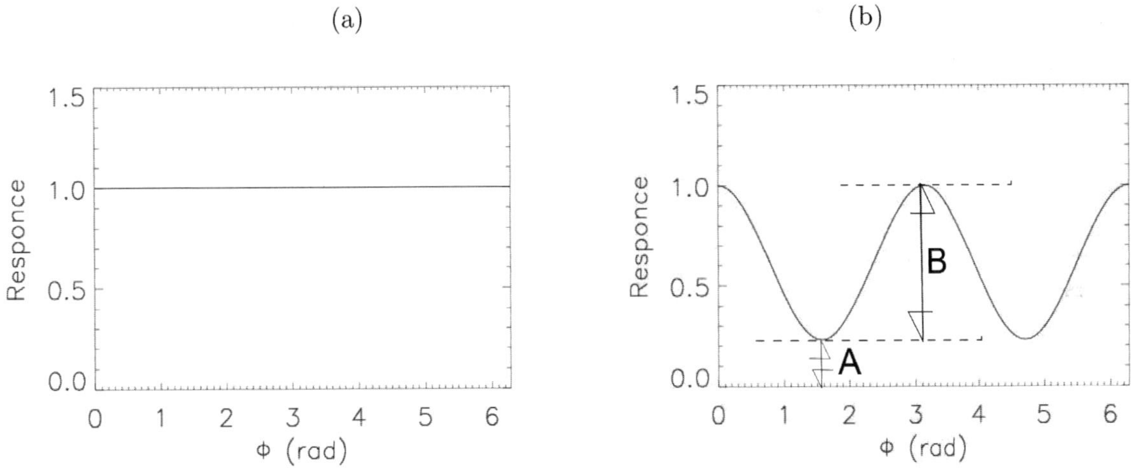

Figure 1. Representation of a polarimeter response depending on the state of polarization of the detected radiation. On the left (a): the signal obtained from unpolarized radiation is the same for all the analysed angular direction ϕ, because there is no a preferential direction of polarization. On the right (b): the signal obtained from polarized radiation is modulated.

the Compton scattering have differential cross sections dependent on a $\cos^2\phi$ term that links the direction of polarization of the incoming photon to the photoelectron azimuthal emission direction in the first case and to the azimuthal direction of Compton scattering in the latter case. Therefore a modulated pattern arises in the detector response that can be fitted by the function:

$$N(\phi) = A_{\mathrm{P}} + B_{\mathrm{P}} cos^2(\phi - \phi_0) \tag{1}$$

where P denotes the polarization degree. The *modulation factor* is defined for 100% polarized radiation as:

$$\mu = \frac{N^{\max}_{100\%} - N^{\min}_{100\%}}{N^{\max}_{100\%} + N^{\min}_{100\%}} = \frac{B_{100\%}}{2A_{100\%} + B_{100\%}} \tag{2}$$

Such a term is a normalization factor needed to calculate the unknown degree of polarization in a generic case. The radiation whose degree of polarization must be measured will produce a modulated pattern response in the detector which will be fitted by the function of Eq. 1. From the modulation factor and the resulting fitted parameter the polarization degree is evaluated as:

$$P = \frac{1}{\mu}\frac{B_{\mathrm{P}}}{2A_{\mathrm{P}} + B_{\mathrm{P}}} \tag{3}$$

The sensitivity of a polarimeter is quantified by means of the Minimum Detectable Polarization (MDP) which is the minimum degree of polarization which can be detected within a certain confidence level [17]. At the 99% of confidence level for a source dominated observation the MDP is given by:

$$MDP(99\%) \simeq \frac{4.29}{\mu\sqrt{N_{\rm ph}}} \quad \text{if} \quad B \ll R_{\rm source} \tag{4}$$

where $N_{\rm ph}$ is the number of detected photons from the source, B is the background and $R_{\rm source}$ is the instrumental source counts rate and the factor 4.29 is derived assuming a Poissonian noise distribution. To achieve an MDP of 1%, with $\mu = 0.5$ the large number of about $7.36{\cdot}10^5$ photons must be detected from the source.

3. Solar flares X-ray emission

Solar flares are huge releases of energy into the solar atmosphere originating from magnetic reconnection. As a consequence heating of coronal plasma and acceleration of particles take place. Particles precipitate downwards to the lower solar atmosphere, while depositing their energy [18, 19, 20]. Electrons slowing down emit via bremsstrahlung the non-thermal Hard X-Ray (HXR) emission. Some times the Soft X-Ray (SHR) emission produced by thermal heating from the magnetic reconnection site is not negligible contradicting the model according to which the atmospheric heating is due simply to particles slowdown [21, 22, 23]. Therefore solar flares emission is characterized by thermal emission from hot plasma below about 10–15 keV, where there is also a strong emission due to lines up to about 7 keV [24], and non-thermal bremsstrahlung from accelerated electrons above about 15–20 keV. The soft emission comes typically from the flare loop, whereas the hard one arises from the loop foot-points and sometimes also from the magnetic reconnection site. Polarized radiation is expected from flares models of non-thermal emission due to the anisotropic distributions of electrons that are accelerated in ordered magnetic fields [25, 26, 27, 28, 29, 30]. Thermal radiation is not expected to be highly polarized, only a few per cent is expected due to possible anisotropies in the electron distribution function [31]. The backscattered radiation by the low solar atmosphere can induce and modify the polarization properties of reflected photons [32, 33].

4. The Gas Pixel Detector

The Gas Pixel Detector (GPD) has been developed by the INFN and the INAF-IAPS Italian research institutes. It exploit the dependence of the photoelectric differential cross section to the polarization of the absorbed photon. In figure. 2a the scheme of the polarimeter is shown. When an X-ray photon is absorbed in the gas a photoelectron is emitted with higher probability in the direction of oscillation of the electric field of the absorbed photon ($\phi = 0$). It propagates ionizing gas atoms, therefore losing its energy. Electron-ion pairs that are produced along the photoelectron path are drifted and amplified by the Gas Electron Multiplier (GEM) and eventually projected on a fine sub-divided pixel detector located to the opposite site of the entrance window (see figure 2b). The GEM is composed by two metal layers to which a difference of potential is applied for charge multiplication. They are separated by a thin insulator, etched with a regular hexagonal pattern of holes. Low Z He-DME based gas mixtures are suitable to measure polarization in the 2–10 keV energy range, while high Z Ar based ones can extend the energy range of employment up to about 35 keV. The analysis of the projected charge distribution statistical momenta allows to evaluate the projection on the pixel plane of the absorption point (useful to obtain the image of the celestial source) and of the photoelectron emission direction (needed to obtain the modulation curve when many photons are detected). For such an instrument the role of the *analyser* is played by the gas, while the *detector* is the pixel plane used to readout the charge signal. Thanks to the imaging capability of the GPD it can be couple with coded mask aperture or X-ray grazing incidence telescope to obtain the image of the source, allowing to get its X-ray polarization map. Sensitivity estimates have been evaluated [13] for this polarimeter coupled with a coded mask aperture assuming an open area

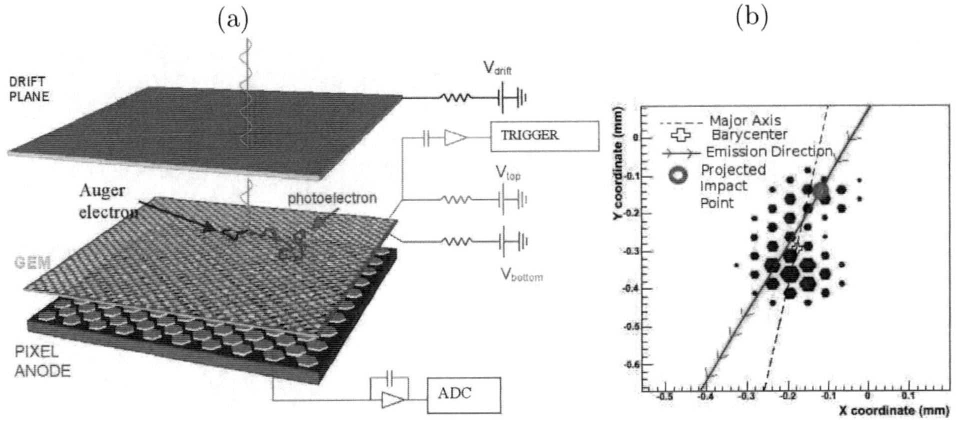

Figure 2. On the left (a): Scheme of the Gas Pixel Detector. On the right (b): 4.5 keV photoelectron charge distribution projected on the pixel plane. The gas mixture is 100% DME at 0.8 bar of pressure.

of 1 cm^2 with respect to the X1.9 flare of 30 August 2002 [34]. An MDP of 5% in the energy band 15–20 keV and 12% in the energy band 20–35 keV were estimated during the peak burst event lasting for 16 s.

5. Compton polarimeter

To measure X-ray polarization at higher energies with respect to the energy range at which the photoelectric absorption is effective, the Compton scattering must be taken into account. Following the Klein-Nishina cross section the scattered photon has a higher probability to be scattered orthogonally to the direction of oscillation of the incoming photon electric vector ($\phi = 90°$). A double phase instrument given by a scatterer made of a low Z material (to have a large scattering efficiency) coupled with a high Z would allow to reduce instrument background taking advantage of the coincidence signals of the scintillation scatterer and the absorber. Therefore it is possible to reduce the low energy threshold down to about 20 keV, where Compton scattering probability in organic scintillators equals the photoelectric absorption. At the present work is in progress and further studies are needed to optimize such an instrument design for solar flares observations.

6. Conclusions

The extension of polarimetry to the hard X-ray energy band allows to study the physics of solar flares. The photoelectric polarimeter Gas Pixel Detector equipped with an Ar based high Z gas mixture can be used to measure the polarization up to about 35 keV. Such an instrument has imaging capabilities and can be coupled with a coded mask aperture or a grazing incidence X-ray telescope to obtain the image of the source. A Compton scattering polarimeter can extend upwards the possibility to measure X-ray polarization to higher energy. If a low energy threshold configuration is chosen, for example by applying coincidence between scatterer and absorber, it is possible to overlap the energy range of this instrument with the GPD Ar based in the energy band 20–35 keV.

References

[1] Weisskopf M C, Cohen G G, Kestenbaum H L, Long K S, Novick R and Wolff R S 1976 ApJ **208** L125
[2] Weisskopf M C, Silver E H, Kestenbaum H L, Long K S and Novick R 1978 ApJ **220** L117

[3] 20120224 `http://heasarc.gsfc.nasa.gov/ docs/ gems/`

[4] Jahoda K 2010 *Proc. of SPIE* (*Society of Photo-Optical Instrumentation Engineers (SPIE) Conference Series* vol 7732)

[5] Black J K, Deines-Jones P, Hill J E, Iwahashi T, Jahoda K, Kaaret P, Kallman T R, Martoff C J, Prieskorn Z, Swank J and Tamagawa T 2010 *Proc. of SPIE* (*Society of Photo-Optical Instrumentation Engineers (SPIE) Conference Series* vol 7732) p 77320X

[6] Soffitta P, Costa E, Muleri F, Campana R, Del Monte E, di Cosimo S, Evangelista Y, Fabiani S, Feroci M, Lazzarotto F, Rubini A, Bellazzini R, Brez A, Minuti M, Omodei N, Pinchera M, Razzano M, Sgrò C, Spandre G, Argan A and Matt G 2010 *Society of Photo-Optical Instrumentation Engineers (SPIE) Conference Series* (*Society of Photo-Optical Instrumentation Engineers (SPIE) Conference Series* vol 7732)

[7] Hosack M, Black J K, Deines-Jones P, Dennis B R, Hill J E, Jahoda K, Shih A Y, Urba C E and Emslie A G 2011 *Experimental Astronomy* **32** 101–125

[8] Kamae T, Andersson V, Arimoto M, Axelsson M, Marini Bettolo C, Björnsson C, Bogaert G, Carlson P, Craig W, Ekeberg T, Engdegård O, Fukazawa Y, Gunji S, Hjalmarsdotter L, Iwan B, Kanai Y, Kataoka J, Kawai N, Kazejev J, Kiss M, Klamra W, Larsson S, Madejski G, Mizuno T, Ng J, Pearce M, Ryde F, Suhonen M, Tajima H, Takahashi H, Takahashi T, Tanaka T, Thurston T, Ueno M, Varner G, Yamamoto K, Yamashita Y, Ylinen T and Yoshida H 2008 *Astroparticle Physics* **30** 72 (*Preprint* `0709.1278`)

[9] Xiong S, Produit N and Wu B 2009 *Nuclear Instruments and Methods in Physics Research A* **606** 552 (*Preprint* `0907.3337`)

[10] Hayashida K 2010 *X-ray and γ-ray polarimetry small-satellite mission PolariS; X-ray Polarimetry: A New Window in Astrophysics by Ronaldo Bellazzini, Enrico Costa, Giorgio Matt and Gianpiero Tagliaferri. Cambridge University Press, 2010. ISBN: 9780521191845, p. 333* p 333

[11] Beilicke M, Baring M G, Barthelmy S, Binns W R, Buckley J, Cowsik R, Dowkontt P, Garson A, Guo Q, Haba Y, Israel M H, Kunieda H, Lee K, Matsumoto H, Miyazawa T, Okajima T, Schnittman J, Tamura K, Tueller J and Krawczynski H 2011 *Society of Photo-Optical Instrumentation Engineers (SPIE) Conference Series* (*Society of Photo-Optical Instrumentation Engineers (SPIE) Conference Series* vol 8145) (*Preprint* `1109.6668`)

[12] Krawczynski H, Garson A, Guo Q, Baring M G, Ghosh P, Beilicke M and Lee K 2011 *Astroparticle Physics* **34** 550–567 (*Preprint* `1012.0321`)

[13] Fabiani S, Costa E, Bellazzini R, Brez A, di Cosimo S, Lazzarotto F, Muleri F, Rubini A, Soffitta P and Spandre G 2012 *Advances in Space Research* **49** 143–149 (*Preprint* `1111.6391`)

[14] Suarez-Garcia E, Hajdas W, Wigger C, Arzner K, Güdel M, Zehnder A and Grigis P 2006 Sol. Phys. **239** 149–172 (*Preprint* `arXiv:astro-ph/0609778`)

[15] Nusinov A A, Kazachevskaya T V and Katyushina V V 2004 *Multi-Wavelength Investigations of Solar Activity* (*IAU Symposium* vol 223) ed Stepanov A V, Benevolenskaya E E and Kosovichev A G pp 475–476

[16] Zhitnik I A, Logachev Y I, Bogomolov A V, Denisov Y I, Kavanosyan S S, Kuznetsov S N, Morozov O V, Myagkova I N, Svertilov S I, Ignat'ev A P, Oparin S N, Pertsov A A and Tindo I P 2006 *Solar System Research* **40** 93–103

[17] Weisskopf M, Elsner R and O'Dell S 2010 *Proceedings of the SPIE,* **7732** 77320E–77320E–5

[18] Brown J C 1971 Sol. Phys. **18** 489–502

[19] Hudson H S 1972 Sol. Phys. **24** 414–428

[20] Lin R P and Hudson H S 1976 Sol. Phys. **50** 153–178

[21] Peres G, Reale F, Serio S and Pallavicini R 1987 ApJ **312** 895–908

[22] Fludra A, Doyle J G, Metcalf T, Lemen J R, Phillips K J H, Culhane J L and Kosugi T 1995 A&A **303** 914

[23] Battaglia M, Fletcher L and Benz A O 2009 A&A **498** 891–900 (*Preprint* `0903.2754`)

[24] Doschek G A 2002 *Stellar Coronae in the Chandra and XMM-NEWTON Era* (*Astronomical Society of the Pacific Conference Series* vol 277) ed Favata F and Drake J J p 89

[25] Brown J C, McClymont A N and McLean I S 1974 Nature **247** 448–449

[26] Langer S H and Petrosian V 1977 ApJ **215** 666–676

[27] Emslie A G and Vlahos L 1980 ApJ **242** 359–373

[28] Leach J and Petrosian V 1983 ApJ **269** 715–727

[29] Zharkova V V, Brown J C and Syniavskii D V 1995 A&A **304** 284

[30] Charikov J E, Guzman A B and Kudryavtsev I V 1996 A&A **308** 924–928

[31] Emslie A G and Brown J C 1980 ApJ **237** 1015–1023

[32] Bai T and Ramaty R 1978 ApJ **219** 705–726

[33] Jeffrey N L S and Kontar E P 2011 A&A **536** A93 (*Preprint* `1110.4993`)

[34] Karlický M, Fárník F and Krucker S 2004 A&A **419** 365–373

Young Researcher Meeting in Rome 2012 IOP Publishing
Journal of Physics: Conference Series **383** (2012) 012014 doi:10.1088/1742-6596/383/1/012014

The birth of Tor Vergata Fabry-Pérot interferometer

L. Giovannelli, F. Berrilli, M. Cocciolo, D. Del Moro, A. Egidi, R. Piazzesi and M. Stangalini

Dipartimento di Fisica, Universitá di Roma *Tor Vergata*, Via della Ricerca Scientifica, I-00133 Rome, Italy

E-mail: giovannelli@roma2.infn.it

Abstract. Fabry-Pérot tunable filters are of great interest in high spectral resolution imaging for both ground-based and space astronomical observations. Major advantages include imaging capabilities and the study of extended astronomical sources, such as the solar photosphere. The high transparency of the instrument allows the high time-resolution necessary for the observation of fast dynamic processes. The prototype here presented has been developed as part of the study for the narrow band channel of the ADAHELI mission. The ADvanced Astronomy for HELIophysics (ADAHELI) is a solar satellite designed to investigate the dynamics of solar atmosphere as part of the Italian Space Agency (ASI) program.

1. Introduction

A Fabry-Pérot (FP) is an instrument based on multiple beam interferometry. It consists of two flat plates which create an optical cavity. The two optical surfaces have a high reflectivity obtained via a dielectric coating; they trap the incoming beam, causing multiple reflections. Interference between the beams depends on the optical path difference (OPD), i. e. the gap between the plates. In this way, the interferometer allows transmission of light at well-defined wavelengths. A spectrometer based on tunable filter is a solution to the modern spectroscopy problem of collecting both spatial and spectral information. Every image taken is nearly monochromatic and it is possible to rapidly change the wavelength of the observation varying the OPD: as a consequence, it is possible to scan absorption lines in an extended range of wavelengths. Jaquinot in 1954 has compared FP and grating spectrometer transparency, having the same instrumental features fixed and operating with the same resolving power [2]. FP spectrometer proved to be the most efficient by a factor of at least 30. This is the reason why the FPs are so attractive for Astrophysics, since the astrophysical sources are usually weak.

If we analyze the spectral response of the instrument, we can see that transparency has several peaks, resembling a comb. The transmission profile depends on the value of the reflectivity R, the reflecting finesse $\mathcal{F}_R$ and the interference order m [1]:

$$\mathcal{F}_R = \frac{\pi\sqrt{R}}{1-R} \tag{1}$$

$$m = \frac{2n}{\lambda} d\cos\theta \tag{2}$$

The order m, once defined the working wavelength, depends on the refracting index n, the angle of incidence θ and the spacing d. The incidence angle of the beam on the instrument is usually set as low as possible, in order to maximize the order, which leaves us with two possible ways to change m: varying either the refracting index n or the spacing d.

The reflecting finesse depends only on the reflectivity R and is set choosing a suitable multilayer dielectric coating. The free spectral range (FSR) and FWHM of the transmission peak are then:

$$FSR = \frac{\lambda}{m} \tag{3}$$

$$FWHM = \frac{FSR}{\mathcal{F}_R} \tag{4}$$

It is evident that the reflecting finesse $\mathcal{F}_R$ and the interference order m are the parameters that define the characteristics of the FP. The FSR is the maximum wavelength range we can scan with a single peak. The FWHM related to the spectral resolution.

We are realizing a prototype interferometer at the Solar Physics Laboratory of University of Rome Tor Vergata. We have chosen the instrument characteristics to match the required performance of a tunable interference filter for solar physics applications, both for ground and space use. The main goal is to produce an instrument suitable for the ADAHELI satellite mission (see section 2). A couple of FP interferometers with an adjustable optical cavity dimension has been proposed as the core of the narrow band channel of the satellite. The prototype is intended to be an important step towards the instrument realization.

2. The ADAHELI mission

ADAHELI is a small-class low-budget satellite mission with the main aim to monitor solar flares and to study the dynamics of different layers of the solar atmosphere [3]. ADAHELI design has completed its Phase-A feasibility study in December 2008, in the framework of ASI (Agenzia Spaziale Italiana) 2007 Small Missions Program. University of Rome Tor Vergata was the scientific institution leader of the project and CGS SpA was the leader industrial partner. ISODY [4], Interferometer for SOlar Dynamics, is the Gregorian telescope of the satellite. It is designed to obtain high resolution spatial, spectral, and temporal polarimetric images of the solar photosphere and chromosphere. The Focal Plane Assembly of the ISODY instrument comprises two visible near-infrared science optical paths or channels: the Narrow Band (NB) and the Broad Band (BB) channels. The principal optical path of the NB channel includes two FP interferometers used in axial-mode and in classic mount and a filter wheel carrying four interference filters used as pre-filters to select one of orders coming from the FPs [5].

3. Prototype Design

The chosen type of FP is a scanning optical cavity with a gap servo control system. The basic idea of the prototype is an instrument capable of maintaining two partially reflecting mirrors at the required distance, parallel within the tolerances. Since the wavelength position of the filter transmission peak is determined by the gap dimension d (see equation 2), this distance must be varied in order to scan a certain spectral range of interest. However, d has to be constant during the instrument acquisition time. In addition, positioning repeatability must be guaranteed in order to obtain the same spectral points in successive measures. To achieve these goals we have used piezoelectric actuators and capacitive sensors. The combination of these two technologies in a servo controlled system grants the necessary stability and repeatability ($\lambda/3000$ over one day or more [6]).

We have designed the opto-mechanics of the laboratory prototype to house a pair of 1" (25.4 mm) partially reflecting mirrors, three micrometers, three piezoelectric actuators and three high sensitivity capacitive sensors. A schematic image of the setup is shown in figure 1.

Journal of Physics: Conference Series **383** (2012) 012014 doi:10.1088/1742-6596/383/1/012014

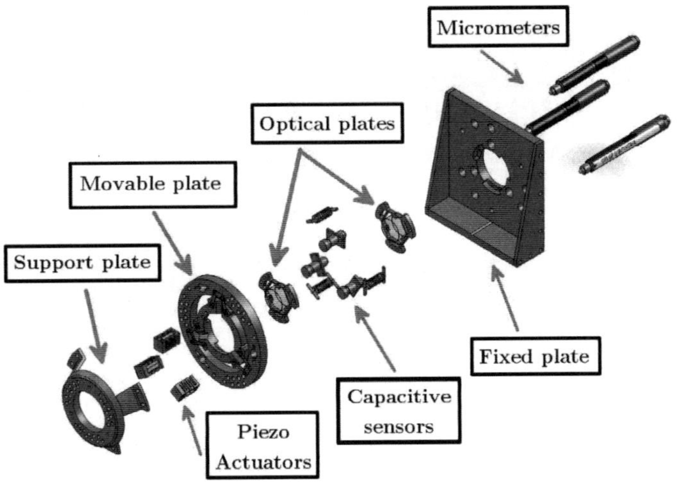

Figure 1. Prototype exploded view.

The maximum allowed travel is 12 mm for micrometers and 15 μm for piezoelectric actuators. The capacitive sensors working distance is 50 μm.

We have designed this prototype to decouple two different movements of the optical surfaces, in order to enable a coarse and fine adjustments of the cavity gap.

This decoupling is achieved by separating the movable plate into two concentric rings. Adjustments are performed either by the three micrometers or the three piezoelectric actuators placed in a 120°-symmetry around the etalon.

The inner ring, which houses the movable optical surface, is controlled by the piezoelectric actuators which can vary the gap size within opto-mechanical specifications. The outer ring is connected to the inner ring by three V-shaped flexures, therefore a micrometer displacement moves both the inner and the outer ring, while a piezoelectric actuator displacement moves the inner ring of the movable plate only.

3.1. Numerical simulation

We developed a software to simulate the performance of the prototype. In order to take into account for non-ideal response of the system, we used an effective finesse.

We simulated three FPs, with three different OPD. The pattern obtained is shown in figure 2 for 3.0 mm (top-left panel), 1.0 mm (top-right panel) and 0.3 mm (bottom-left panel). The broadening of the transmission profile and the lowering in transparency for the real case (dashed line) is significant and shown in the previously mentioned panels. The simulation aim is to decide what value of the OPD makes the instrument more suitable to scan the H-alpha line, taking into account the available prefilter.

In a trade-off between the FSR and FWHM set by the optical gap, and using equation 4, the 1.0 mm value is accepted for the tests.

3.2. Piezo tests

Fine adjustments of the movable optical plate is achieved using piezoelectric actuators. Piezo actuators can perform sub-nanometer moves, have no rotating or sliding parts that cause friction and dissipate virtually no power in static operation. They have a nominal displacement of 15 μm under an applied voltage of 100 V and have the possibility to reach nanometric precision. They

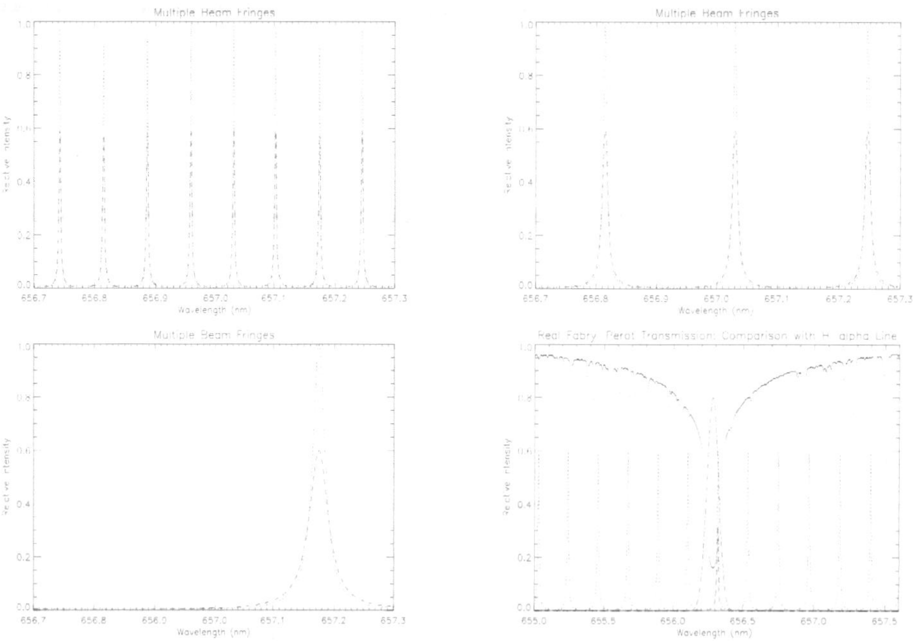

Figure 2. FP numerical simulations. Dotted: ideal FP profile; dashed: effective FP profile. From upper left to bottom right the OPDs are: 3 mm, 1 mm, 0.3 mm, 1 mm together with H_α absorption line and prefilter profile (dot-dashed).

are operated using an analog driving signal; the system is computer operated via a Digital to Analog Converter (DAC) that has a 16-bit resolution, resulting in 0.15 mV minimum output voltage step. After the signal processing by the controller, this results in a 0.44 nm minimum step of the piezo actuator.

Unfortunately, the piezoelectric actuator response to the signal is non-linear and follows a hysteresis cycle. This issue can be overcome using a closed loop control system, which involves a high precision distance sensor. In this case, we have used a capacitive sensor with a static resolution $< 1\ pm$. The capacitive sensor has a working range centered at a distance of 50 μm from the target. The piezo actuator has been proven to be able to operate in low pressure and low temperature compatible with space environments (2 K and 0.5 bar).

In order to test the reliability of the opto-mechanical actuator, both in closed and open loop operation mode, we realized a device to test the actuator in a controlled situation on the optical bench and developed the dedicated software.

We obtained open loop measures using 1 V steps from $-1\ V$ to 10 V as low-voltage driving signal. The hysteresis cycle of the piezo is evident (see the left panel of figure 3): bottom points refers to piezo elongation, while top points refers to piezo retreat. Maximum elongation is $14.1 \pm 0.2\ \mu m$, which is near the nominal value of 15 μm.

We also obtained closed loop measures using a reduced control voltage range, still able to cover the full range of piezo extension. 0.5 V steps from 1.5 V to 5.5 V were used as low-voltage driving signal.

The response is linearized with respect to the open loop data as it is evident from the linear fit and related residuals (middle and right panel of figure 3).

Also the maximum extension of the piezo is $14.8 \pm 0.2\ \mu m$ in closed loop mode.

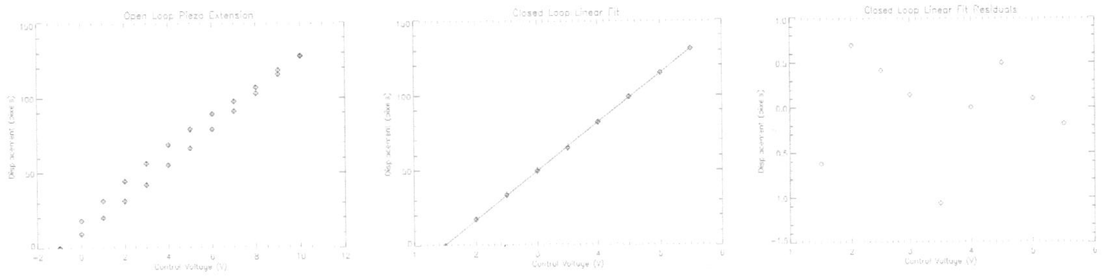

Figure 3. Piezo-actuators elongation data, from left to right: open loop hysteresis, closed loop with linear fit and related residuals.

3.3. Initial assembly phase

Prototype assembly has started with the production of the first pieces. We chose stainless steel for the construction of all the mechanical parts. Stainless steel has been selected mainly for its stiffness properties, low cost and availability. The two concentric rings of the movable plate have been realized with a maximum surface error $< 40\ \mu m$ over the whole surface. Images of the plate are provided in figure 4.

Figure 4. Movable plate: inner and outer rings (left), rings and support plate (right).

4. Conclusions and Future Developments

In this work, we have presented the design of a laboratory prototype Capacitance Stabilized Fabry-Pérot interferometer. First, we introduced the basic FP theory and presented the results of a FP numerical model. From these results, we designed a FP prototype optimized for the scan of the H_α line, but also suitable for the whole visible range. Along with its implementation, we have presented the mechanical, electronic and optical characteristics of the prototype in details and some of the tests we performed on its basic components.

In the near future, we plan to execute the optical and spectral qualification of this prototype.

Furthermore, we are working on a space implementation of the FP in order to start space qualification tests. The design has been update to match the more demanding operating environment requirements, such as the launch stresses and the large thermal variations.

References

[1] Wolf M Born E 1999 7th ediction Principles of Optics: Electromagnetic Theory of Propagation, Interference and Diffraction of Light *Cambridge University Press*
[2] Hernandez G 1988 Fabry-Perot interferometers *Cambridge University Press*

[3] ADAHELI Team *et al* 2010 *Advances in Space Research* **45** 1191
[4] Greco V Cavallini F Berrilli F 2010 *Society of Photo-Optical Instrumentation Engineers (SPIE) Conference Series* **7731**
[5] Berrilli F *et al* 2011 *Society of Photo-Optical Instrumentation Engineers (SPIE) Conference Series* **8148**
[6] Hicks T R Ready N K and Atherton P D 1984 *Journal of Physics E Scientific Instruments* **17** 49

Structural, electronic and optical properties of the two isomers of Si(111)2x1

C. Violante[1], A. Mosca Conte[1] and O. Pulci[1,2]

[1]ETSF, MIFP, Dipartimento di Fisica, Università di Roma Tor Vergata, Via della Ricerca Scientifica 1, 00133 Rome, Italy

[2]Istituto di Struttura della Materia, Consiglio Nazionale delle Ricerche, Via del Fosso del Cavaliere 100, I-00133 Rome, Italy

E-mail: claudia.violante@roma2.infn.it

Abstract. The Si(111)2x1 is one of the most studied surfaces. Its reconstruction is described by the Pandey model with a buckling of the topmost atoms. With relation to the sign of the buckling, there are two slightly different geometric structures (isomers), conventionally named positive buckling and negative buckling. STM measurements suggest that the positive buckling isomer is the stable configuration at room temperature, but a recent work, involving STS measurements, has shown the coexistence of both the isomers, at very low temperature, for highly n-doped Si(111)2x1 specimens. There is hence the necessity to deepen the study of the negative buckling isomer, almost completely neglected in literature especially for what concerns its optical properties. In this work we have studied the structural, electronic and optical properties of both the isomers of Si(111)2x1 within the Density Functional Theory in the Local Density Approximation. Our results show that the response to light of the two isomers is sizearly different; Reflectance Anisotropy Spectroscopy would be the right experimental tool to investigate the coexistence of the two isomers.

1. Introduction

The (111) surfaces are the natural cleavage planes of Si and the 2x1 is the reconstruction observed at low temperature after cleavage in Ultra High Vacuum. Si(111)2x1 is one of the most investigated surfaces and its atomic structure is well understood in terms of the Pandey model [1]. According to the latter, π-bonded atoms chains are present in the first atomic layer. Two adjacent atoms of the chain can tilt, with respect to each other, in two ways and so two different isomers of the surface, conventionally named positive buckling and negative buckling, can be distinguished (see **Figure 1**).

The geometry of the two isomers is very similar but only the positive buckling structure is observed by STM measurements [2] at room temperature. However, for highly n-doped Si(111)2x1 specimens, at very low temperature, the coexistence of the two structures has recently been shown by a comparison between experimental and theoretical STS [3]. A deeper understanding and the final proof of this coexistence can be obtained by the study of the optical properties of these two configurations. There is hence the need to investigate more deeply the negative buckling structure, in particular for what concerns the optical properties. In this work we have studied the geometry, electronic band structure and Reflectance Anisotropy Spectra of both the isomers by Density Functional Theory (DFT) ab-initio calculations.

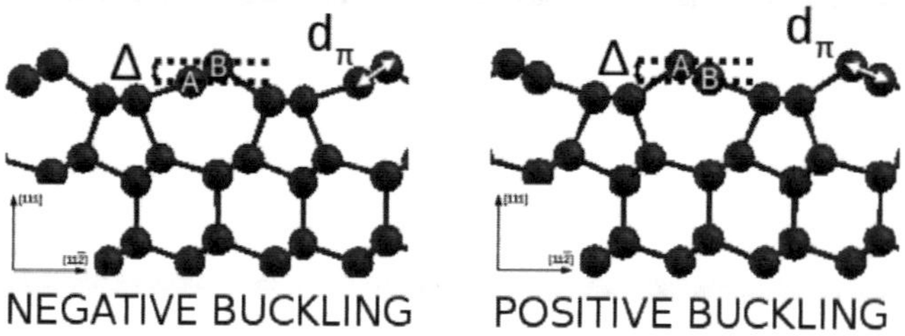

Figure 1. side view of Si(111)2x1 negative and positive buckling reconstructions.

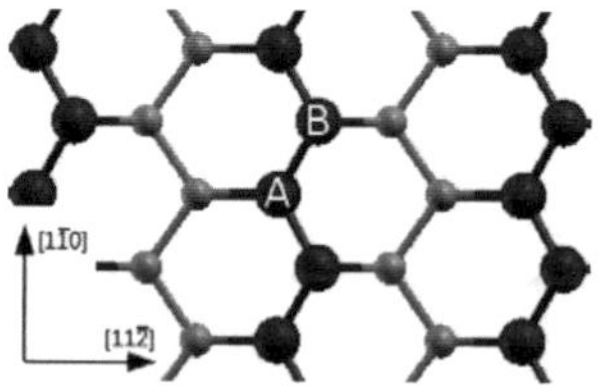

Figure 2. top view of Si(111)2x1 negative and positive buckling. The bigger atoms are the ones in the first atomic layer. The distance between atom A and atom B is the bond length along the Pandey chain while the difference in height between atom A and atom B is the buckling amplitude.

2. Methods

We have performed DFT calculations in the Local Density Approximation (LDA), using the codes of *Quantum Espresso* package [4]. The surfaces have been modelled using the supercell method, employing slabs of 24 atoms separated by 7.42 Å of vacuum. We have used different values of cut-off energies: 30 Ry for geometry, 40 Ry for structural properties and 20 Ry for optical spectra. For what concerns the optical properties, we have focused on the RAS (Reflectance Anisotropy Spectroscopy) spectra. RAS [5][6][7] is a powerful and efficient experimental technique widely used to study and characterize surfaces. The RAS signal measures the difference of reflectivity for light polarized along two perpendicular directions (x and y) on the plane of the surface:

$$\frac{\Delta R}{R} = \frac{R_y - R_x}{R}$$

and it corresponds to the difference of the reflectivity corrections to the Fresnel (unperturbate) case for the two different light polarizations. For cubic crystals, the RAS signal comes from the surface anisotropy, since the bulk is optically isotropic. In the case of the supercell method, there is a simple theoretical expression for the RAS spectrum [8], involving the half-slab polarizability α_{ii}^{hs}:

$$\frac{\Delta R}{R} = \frac{4\omega d}{c} Im \frac{4\pi\alpha_{ii}^{hs}}{\epsilon_b - 1}$$

$$\alpha_{ii}^{hs(2)}(\omega) = \frac{\pi e^2}{m^2\omega^2 Ad} \sum_{\vec{k},v,c} |p_{vc}^i(\vec{k})|^2 \delta(E_c(\vec{k}) - E_v(\vec{k}) - \hbar\omega)$$

where d is the half-slab thickness, A is the 2D cell area, ϵ_b is the bulk dielectric function, i is the direction of light polarization and the sum in the second equation runs over the k points in the Irreducible Brillouin Zone (IBZ) and over conduction (c) and valence (v) states. We have used 231 k points in IBZ and 800 total bands (48 filled states).

3. Results

3.1. Structural properties

Ground state configurations of both the isomers of Si(111)2x1 have been obtained minimizing DFT total energy. Here we show the most important geometric parameters, that are the buckling amplitude Δ and the bond length d_π along the π-chains (defined in **Figure 1** and **2**):

$$\Delta^{pb} = +0.53 \text{ Å and } d_\pi^{pb} = 2.27 \text{ Å, for the positive buckling}$$

$$\Delta^{nb} = -0.59 \text{ Å and } d_\pi^{nb} = 2.27 \text{ Å, for the negative buckling}$$

Our results are in good agreement with literature [2][3][9][10][11][12].

As we can see in **Figure 3**, the two isomers of Si(111)2x1 are both local stable energy minima. The energetic barrier, which separates the two stable configurations, can be roughly estimated by switching from one configuration to the other simply tilting the buckled atoms of the Pandey chain with respect to each other. We find that 0.06 eV/(2x1) is an upper limit for the energy barrier.

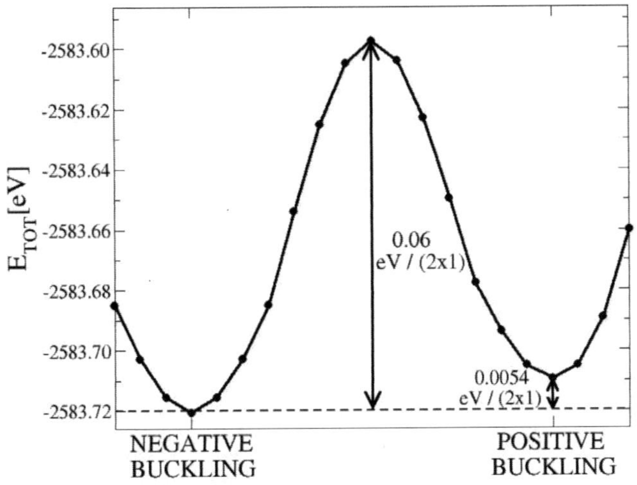

Figure 3. Energy path from the negative buckling structure to the positive buckling configuration.

In our calculations, the negative buckling is the most stable configuration, being favored by about 5.4 meV/(2x1) with respect to the positive buckling. Anyway the energy difference is so small that we can assume the structures nearly degenerate within DFT accuracy, as well known in literature [13][14].

3.2. Electronic properties

The calculated energy gaps between surface states at the surface high simmetry points of the Irriducible Brillouin Zone are reported in **Table 1**. We find a good agreement with previous DFT results [14][15].

Table 1. Si(111)2x1 - DFT Electronic gaps [eV]

Positive buckling			Negative buckling		
$\overline{J}$	$\overline{K}$	$\overline{J'}$	$\overline{J}$	$\overline{K}$	$\overline{J'}$
0.35	0.62	1.67	0.19	0.35	1.54

The band structure along the $\overline{JK}$ line for both the isomers is shown in **Figure 4**. The $\overline{JK}$ direction in the reciprocal space corresponds, in real space, to the direction perpendicular to the

Pandey chains: the little dispersion that can be observed along that line is a direct consequence of the weak interaction between two adjacent π-chains (that are nearly 6.6 Å distant). The maximum value of the filled surface band is at $\overline{J}$ for the positive buckling and between $\frac{\overline{JK}}{2}$ and $\overline{K}$ for the negative buckling; the minimum value of the empty surface band is between $\overline{J}$ and $\frac{\overline{JK}}{2}$ for both the isomers.

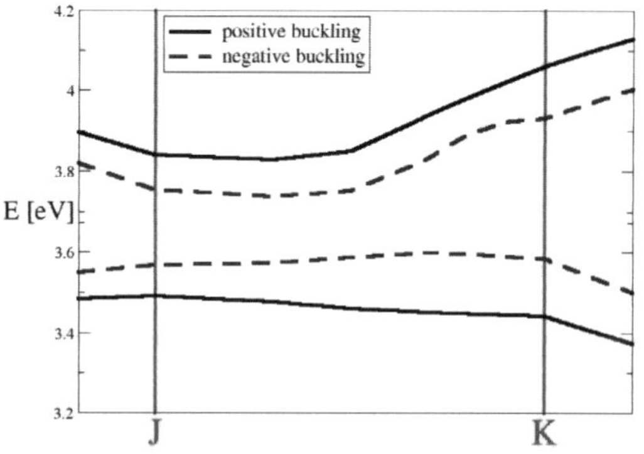

Figure 4. Surface band structure for the positive buckling (solid black line) and the negative buckling (dashed blue line).

For all the k points along $\overline{JK}$ direction the highest occupied states are located on the upper atoms of the Pandey chain, while the lowest empty states are located on the lower atoms of the π-chain. As an example we show the squared modulus plot of the wavefunctions at k-point $\overline{J}$ for the negative buckling isomer (see **Figure 5** and **Figure 6**).

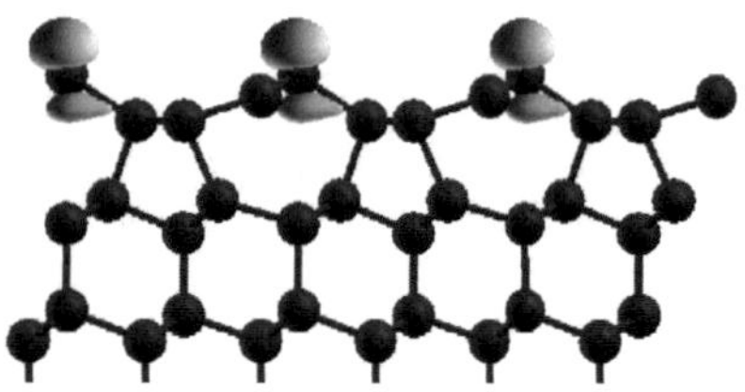

Figure 5. Squared modulus plot of the DFT wavefunction of the highest occupied surface state at $\overline{J}$ point for the Si(111)2x1 negative buckling.

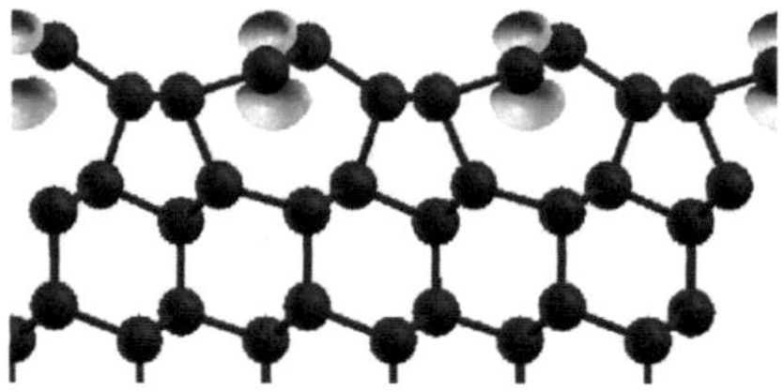

Figure 6. Squared modulus plot of the DFT wavefunction of the lowest unoccupied surface state at $\overline{J}$ point for the Si(111)2x1 negative buckling.

3.3. Optical properties

RAS spectra have been calculated, within DFT-LDA (in the indipendent particle approximation), up to 6 eV and, as we can see in **Figure 7**, there are deep differences in the shape, the intensity and the peak position for the two isomers.

The Si(111)2x1 negative buckling presents a RAS peak (0.22 eV) at lower energies with respect to the peak of the positive buckling (0.48 eV), and this is related to the the fact that the electronic gaps of the positive buckling are larger than the electronic gaps of the negative buckling. The RAS peak of the negative buckling is the most intense (the intensity ratio between the two isomer is nearly 1.6): as we can see in **Figure 4**, the surface bands are more parallel for the negative buckling structure, thus enhancing the optical response.

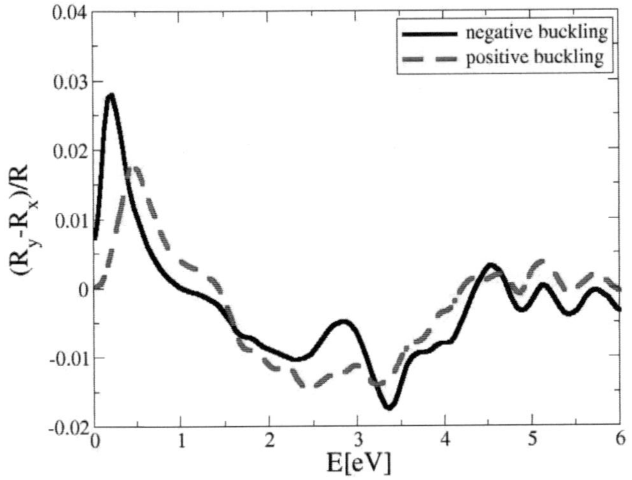

Figure 7. RAS spectra of Si(111)2x1 positive buckling (dashed red line) and negative buckling (solid black line) calculated within DFT-LDA. y is the $[1\bar{1}0]$ direction, parallel to the Pandey chains and x is the $[11\bar{2}]$ direction, perpendicular to the π-chains. A broadening of 0.05 eV has been included.

The results obtained for the Si(111)2x1 positive buckling are in good agreement with previous DFT-LDA RAS spectra [15]. The difference of about 0.26 eV in the energy positions, found in the RAS peak, suggests that experiments would be able to distinguish the two isomers.

4. Conclusions

The isomers of Si(111)2x1 are both energetic stable local minima, almost degenerate in energy and separated by an energetic barrier. A slight difference in the geometry of the two configurations (they are inequivalent with respect to the third atomic layer) leads to deep differences in the electronic and, consequently, in the optical properties. In particular we have found that, within DFT-LDA approximation, the RAS peak of the negative buckling isomer is more intense and is redshifted by $0.2 \div 0.3$ eV with respect to the positive buckling. In order to compare in a more quantitative way with experimental results, many body and excitonic effects must be included in the calculations, and this is the content of a forthcoming work.

Acknowledgments

We would like to thank CINECA and ENEA-CRESCO supercomputer centers for cpu-time. We acknowledge financial support within the EUFP7 CLERMONT4 project (GA235114).

References

[1] K.C. Pandey, Phys. Rev. Lett. **47**, 1913 (1981)
[2] S. Nie et al., J. Vac. Sci. Technol. A **22(4)**, 1671 (2004)
[3] G. Bussetti et al., Phys. Rev. Lett. **106**, 067601 (2011)
[4] http://www.quantum-espresso.org/
[5] "Optical Characterization of Epitaxial Semiconductor Layers", G. Bauer, W. Richter, Springer-Verlag, 1996
[6] D.E. Aspnes et al., J. Vac. Sci. Technol. A **6**, 1327 (1988)
[7] P. Weightman et al., Rep. Prog Phys. **68**, 1251 (2005)
[8] F. Manghi et al., Phys. Rev. B **39**, 13005 (1989)
[9] C.H. Patterson et al., Phys. Rev. B 84, 155314 (2011)
[10] F. Bechstedt et al., Phys. Rev. Lett. **87**, 16103 (2001)
[11] M. Rohlfing and S. G. Louie, Phys. Rev. Lett. **83**, 856 (1999)
[12] G. Xu et al, Phys. Rev. B **70**, 045307 (2004)
[13] F. Ancillotto et al., Phys. Rev. Lett. **65**, 3148 (1990)
[14] M. Zitzlsperger et al. , Surf. Sci. **108**, 377 (1997)
[15] O. Pulci et al., Phys. Rev. B **82**, 205319 (2010)

Journal of Physics: Conference Series **383** (2012) 012016 doi:10.1088/1742-6596/383/1/012016

Monte Carlo simulations of single polymer force-extension relations

F. Manca[1], S. Giordano[2], P. L. Palla[2,4], F. Cleri[2,4] and L. Colombo[1,3]

[1]Dipartimento di Fisica, Università di Cagliari, Cittadella Universitaria, 09042 Monserrato, Cagliari, Italy

[2]Institute Electronique, Microélectronique et Nanotechnologie (IEMN), UMR CNRS 8520, Avenue Poincaré, BP 60069, 59652 Villeneuve d'Ascq, France

[3]IOM-CNR (Unità Cagliari), Cittadella Universitaria, 09042 Monserrato, Cagliari, Italy

[4]Université de Sciences et Technologies de Lille (USTL), Cité Scientifique, 59655 Villeneuve d'Ascq, France

E-mail: fabio.manca@dsf.unica.it, luciano.colombo@dsf.unica.it

Abstract. We present Monte Carlo simulations for studying the statistical mechanics of arbitrarily long single molecules under stretching. In many cases in which the thermodynamic limit is not satisfied, different statistical ensembles yield different macroscopic force-displacement curves. In this work we provide a description of the Monte Carlo simulations and discuss in details the assumptions adopted.

1. Introduction

In biological systems many important molecules are polymer chains. So far, different mechanical experiments on single molecules (e.g., stretching) provided a crucial elucidation about the molecoular behaviour of several biomolecular processes [1]. Single molecules tecniques may include mechanical and optical methods, such as laser optical tweezers (LOTs) [2], magnetic tweezers (MTs) [3] or atomic force microscope (AFM) [4]. The importance of understanding the force-extension relationship for macromolecules has attracted the attention of scientists, who produced several models and relationships to explain the experimental results. The earlier experiments on double-stranded DNA chains showed results in partial agreement with the freely jointed chain (FJC) model [5] but in very good agreement with the worm-like chain (WLC) model [6, 7].

The analytical developments of the FJC and WLC models work under the assumption of the finite, but large enough, contour length. This approximation is related to the concept of thermodynamic limit. In fact, the standard rules of equilibrium thermodynamics may not apply to experiments on individual, short-length polymer molecules. In such a case the results may depend on the boundary conditions imposed for stretching the polymer: a fixed end-to-end distance, pertinent to the Helmholtz ensemble, or a fixed force applied at one or both ends, rather representing the Gibbs ensemble. In this paper we consider the mechanical stretching response of extensible polymer chains of arbitrary length. To this aim we develop Metropolis Monte Carlo simulations performing the typical experiment in which the elasticity of single polymer is probed with an AFM tip (see Fig.1, left panel) and we quantitatively evaluate the stretching response differences between the Helmholtz and Gibbs ensembles (Fig.1, right panel).

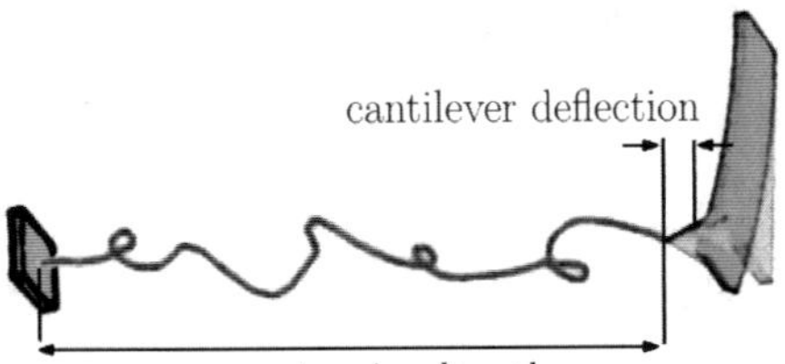

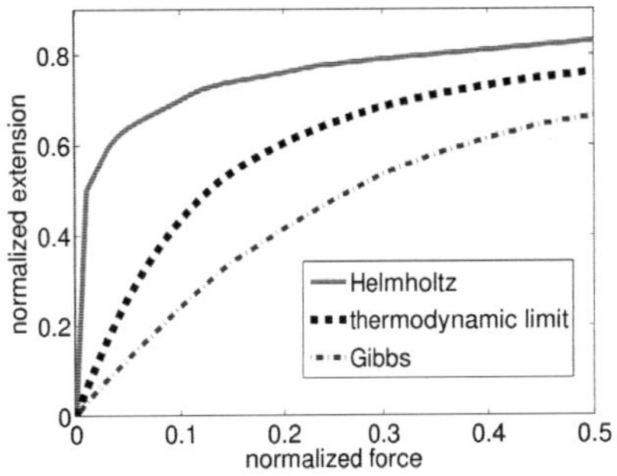

Figure 1. Left panel: a typical AFM experiment to measure the force-extension curve of a single molecule. Right panel: comparison between the normalized extensions in the Helmholtz and Gibbs ensembles for a WLC chain with $N = 10$ monomers.

2. Thermodynamics of polymer chains

Let us consider a chain of monomers in a long molecule, each monomer representing a group of atoms or molecules along the polymer backbone. The classical dynamics of monomers is described by the set of positions $\vec{r}_i$ $(i = 1, ..., N)$ and momenta $\vec{p}_i$ $(i = 1, ..., N)$. We assume that one terminal monomer is fixed at position $\vec{r}_0 \equiv (0, 0, 0)$ and that monomers interact through a harmonic potential. The dynamics of the system is described by the Hamiltonian

$$h_0(\vec{r}_1, .., \vec{r}_N, \vec{p}_1, .., \vec{p}_N) = \sum_{i=1}^{N} \frac{\vec{p}_i \cdot \vec{p}_i}{2m} + \frac{1}{2}k \sum_{i=0}^{N-1} (\|\vec{r}_{i+1} - \vec{r}_i\| - l)^2 . \tag{1}$$

We consider this system to be in thermal equilibrium with a reservoir at temperature T and, therefore, its statistical properties are described by the canonical ensemble distribution $\rho(q,p) = \frac{1}{Z} e^{-\frac{h_0(q,p)}{k_B T}}$ where we have introduced the canonical variables $q = (\vec{r}_1, ..., \vec{r}_N)$ and $p = (\vec{p}_1, ..., \vec{p}_N)$, the Boltzmann constant k_B and the partition function Z.

2.1. Polymer chain with fixed end-to-end distance

By setting a given end-to-end distance, positions $\vec{r}_0$ and $\vec{r}_N$ are fixed and we can use the following reduced Hamiltonian

$$h(\vec{r}_1, .., \vec{r}_{N-1}, \vec{p}_1, .., \vec{p}_{N-1}, \vec{r}) = h_0(\vec{r}_1, .., \vec{r}_{N-1}, \vec{r}_N = \vec{r}, \vec{p}_1, .., \vec{p}_{N-1}, \vec{p}_N = 0). \tag{2}$$

In this case the microscopic variables are defined as $q = (\vec{r}_1, .., \vec{r}_{N-1})$ and $p = (\vec{p}_1, .., \vec{p}_{N-1})$, in terms of which the system partition function is written

$$Z_{\vec{r}}(\vec{r}, T) = \iint_{\Gamma_{N-1}} e^{-\frac{h(q,p,\vec{r})}{k_B T}} dq dp, \tag{3}$$

where $\Gamma_{N-1} = \Re^{6(N-1)}$. The net force exerted on the monomer at position $\vec{r}$ by the remaining monomers is by definition $-\frac{\partial h}{\partial \vec{r}}$. Such a force can be used to define the mechanical constitutive equation of the chain. In fact, the force exerted on the system (from the outside) in order to keep fixed the last monomer is $\vec{f} = \langle \frac{\partial h}{\partial \vec{r}} \rangle$, thus providing the statistical nonlinear generalization of Hooke's law for the chain

$$\vec{f}(\vec{r}, T) = -k_B T \frac{\partial}{\partial \vec{r}} \log Z_{\vec{r}} = \frac{\partial F(\vec{r}, T)}{\partial \vec{r}}, \tag{4}$$

where $F(\vec{r}, T) = -k_B T \log Z_{\vec{r}}$ is the Helmholtz free energy.

Journal of Physics: Conference Series **383** (2012) 012016 doi:10.1088/1742-6596/383/1/012016

2.2. Polymer chain under constant load

We next assume that a given force $\vec{f}$ is applied to the terminal monomer at $\vec{r}_N$, while the end-to-end distance is free to fluctuate. This external force is described by an additional potential energy term given by $-\vec{f}\cdot\vec{r}_N$ since $-\frac{\partial}{\partial \vec{r}_N}(-\vec{f}\cdot\vec{r}_N) = \vec{f}$. Therefore, the system is described by the following augmented Hamiltonian

$$\widetilde{h}(\vec{r}_1, ..., \vec{r}_N, \vec{p}_1, ..., \vec{p}_N, \vec{f}) = h_0(\vec{r}_1, ..., \vec{r}_N, \vec{p}_1, ..., \vec{p}_N) - \vec{f}\cdot\vec{r}_N, \tag{5}$$

where $q = (\vec{r}_1, ..., \vec{r}_N)$ and $p = (\vec{p}_1, ..., \vec{p}_N)$ are the microscopic variables and $\vec{f}$ acts as a macroscopic variable. The ensemble partition function is now given by

$$Z_{\vec{f}}(\vec{f}, T) = \iint_{\Gamma_N} e^{-\frac{\widetilde{h}(q,p,\vec{f})}{k_B T}} dq dp, \tag{6}$$

where $\Gamma_N = \Re^{6N}$. We observe that $\frac{\partial \widetilde{h}}{\partial \vec{f}} = -\vec{r}_N$ and we calculate the mean position of the last monomer of the chain through the average value $\vec{r} = \langle \vec{r}_N \rangle$ or, more explicitly, $\vec{r} = -\langle \frac{\partial \widetilde{h}}{\partial \vec{f}} \rangle$. This constitutive equation can also be expressed in terms of the partition function $Z_{\vec{f}}$. In fact, by differentiating Eq.(6) with respect to $\vec{f}$, we get

$$\vec{r}(\vec{f}, T) = k_B T \frac{\partial}{\partial \vec{f}} \log Z_{\vec{f}} = -\frac{\partial G(\vec{f}, T)}{\partial \vec{f}}, \tag{7}$$

where $G(\vec{f}, T) = -k_B T \log Z_{\vec{f}}$ is the Gibbs free energy. It is possible to prove that the Helmholtz and Gibbs ensembles provide the same constitutive equation for large systems ($N \to \infty$) [8].

3. Flexible polymer with elastic bonds

We firstly consider a model in which each bond is represented by a harmonic spring with finite extension, while no potential is acting on bending or torsional degrees of freedom. Each spring is defined by the potential energy $V(x) = (1/2)k(x - l)^2$ for $x \in (l - \Delta, l + \Sigma)$ where k is the spring constant, l is the equilibrium bond length and x is the actual extension of the bond. We start with the Gibbs ensemble considering the augmented Hamiltonian in Eq.(5). In this case the determination of the spherically-symmetric partition function $Z_f(f, T)$ can be made in closed form (Ref. [8]) and directly provides the (scalar) constitutive equation $r(f) = k_B T \partial \log Z_f(f, T)/\partial f$. Similarly, using the proper partition function for the Helmholtz ensemble $Z_r(r, T)$, we obtain the scalar constitutive equation $f(r) = -k_B T \partial \log Z_r(r, T)/\partial r$ representing the elastic behaviour in the Helmholtz ensemble which depends on the number of monomers N (Ref. [8]). We expect a family of curves approaching the Gibbs solution for large N (thermodynamic limit).

4. Monte Carlo simulations

The Monte Carlo approach simulates the stretching of the polymer under a force provided by a cantilever mimiking the loading by an atomic force microscope. The device is characterized by its own effective elastic constant k_c, which is coupled in series to the chain of k domain springs. In the limit of a soft device, $k_c/k \to 0$, the statistics of the coupled system reduces to the Gibbs ensemble for the isolated molecule fluctuating under a constant force. On the other hand, for a very stiff device, $k_c/k \to \infty$, one recovers the Helmholtz ensemble for the isolated molecule held at a fixed extension by the fluctuating force [9]. In details, we adopted $k_c/k = 10^{-3}$ (soft cantilever) and $k_c/k = 10^2$ (stiff cantilever).

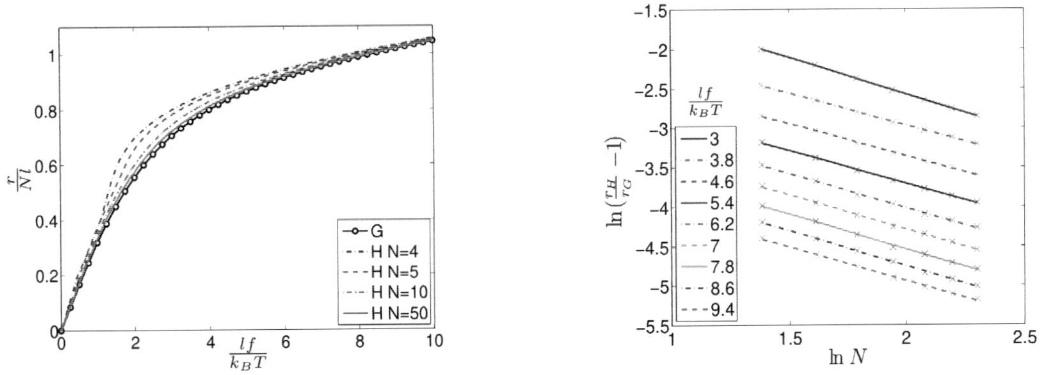

Figure 2. Left panel: constitutive relation provided by Monte Carlo simulations for a FJC model with elastic bonds between the monomers, both under Helmholtz (H) and Gibbs (G) boundary conditions. Right panel: comparison between the elongations r_H and r_G for the FJC model with elastic bonds between the monomers.

The initial state of the chain is defined by a set of randomly chosen positions for the monomers. The displacement extent $\delta\vec{r_i}$ of the positions vectors governs the magnitude of the trial move and the overall efficiency of the configurational space sampling. However, while a larger $\delta\vec{r_i}$ could speed up the search for the minimum, a too large $\delta\vec{r_i}$ leads to a high rejection frequency. Therefore, we analysed several runs in order to optimize its value. We avoided the dynamical adjustment of this parameter since this approach can violate the detailed balance [10].

In Fig.2 (left panel) we report our Monte Carlo results for the FJC model with elastic bonds in both statistical mechanics ensembles. In particular, we selected the value $k = 10\ k_BT/(\text{nm})^2$ at $T = 293$ K. While we observe a single curve for the Gibbs ensemble we get different elastic response curves for the Helmholtz ensemble with polymer length $N = 4, 5, 10$ and 50. It is interesting to note that the Helmholtz curves approach the Gibbs one for large N, as expected for the convergence to the thermodynamic limit. In order to better characterize the convergence toward the thermodynamic limit, we investigate the ratio between the elongation $r_H(N) = r/(Nl)$ calculated in the Helmholtz ensemble and the elongation $r_G = r/(Nl)$ for the Gibbs ensemble. Fig.2 (right panel) proves that Monte Carlo simulations are nicely fitted by the power law $r_H(N)/r_G = 1 + a/N^\alpha$ where a and α are fitting parameters. Each curve provides the same scaling exponent $\alpha = 0.80 \pm 0.05$. Therefore, we argue that the convergence to the thermodynamic limit of the FJC model is quantitatively controlled by a unique scaling exponent.

5. Semiflexible polymer with elastic bonds

We now extend our model to a semiflexible chain which incorporates elastic bonds into a discrete version of the WLC. The augmented Hamiltonian for the Gibbs ensemble is given by Eq.(2) with the additional term $\frac{1}{2}\kappa\sum_{i=1}^{N-1}\left(\vec{t}_{i+1}-\vec{t_i}\right)^2$ where κ is the bending modulus and $\vec{t_i}$ is the unit vector collinear with the i-th bond. The corresponding reduced Hamiltonian for the Helmholtz ensemble is given by Eq.(5) with the same additional term. In Fig.3 (left panel) we report the Monte Carlo results for the WLC model with elastic bonds. We adopted the value $\kappa = 10\ k_BT$ for the bending modulus. This value is comparable to that of polymer chains of biological interest (for example for DNA $\kappa = 15\ k_BT$[11]). We consider again the data in Fig.3 (left panel) and the ratio $r_H(N)/r_G(N)$ (Fig.3, right panel) at a given fixed value of the normalized force. As before Monte Carlo simulations are nicely fitted by the power law $r_H(N)/r_G(N) = 1+a/N^\alpha$

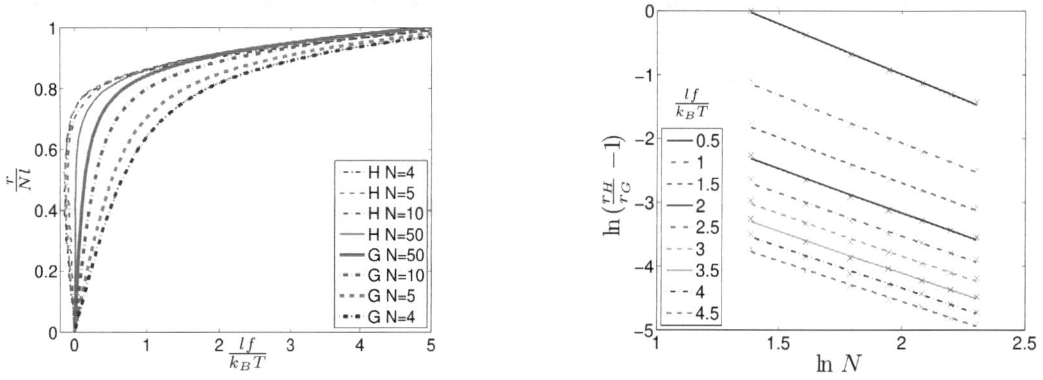

Figure 3. Left panel: constitutive relation provided by Monte Carlo simulations for the WLC model with elastic bonds between the monomers, both under Helmholtz (H) and Gibbs (G) boundary conditions. Right panel: comparison between the elongations r_H and r_G for the WLC model with elastic bonds between the monomers.

where $\alpha = 1.40 \pm 0.05$. Again, the convergence to the thermodynamic limit is controlled by a unique scaling exponent.

6. Conclusions

In this work we investigated how the force-extension curve of a model polymer chain is affected by the loading protocol which can typically be fixed-ends or fixed-force. We showed how such macroscopic boundary conditions can be formulated within the Helmholtz and the Gibbs ensembles of the statistical mechanics. We adopted flexible and semiflexible polymer models, with extensible bonds. In all cases here investigated (FJC and WLC with extensible bonds) we found that the convergence to the thermodynamic limit is well described by suitable power laws with well defined scaling exponents.

Acknowledgments

One of us (L.C.) acknowledge financial support by Regional Government of Sardinia under project M4C (Progetti di Ricerca di Base).

References

[1] F. Ritort 2006, J. Phys.: Condens. Matter **18**, R531
[2] A. Ashkin 1997, Proc. Natl Acad. Sci. **94**, 4853
[3] C. Gosse and V. Croquette 2002, Biophys. J. **82**, 3314
[4] D. M. Czajkowsky and Z. Shao 1998, FEBS Lett. **430**, 51
[5] J. M. Huguet et al 2010, Proc. Nat. Ac. Sci. (PNAS) **107**, 15341
[6] S. B. Smith, L. Finzi, C. Bustamante 1992, Science **258**, 1122
[7] J. F. Marko and E. D. Siggia 1995, Macromolecules **28**, 8759
[8] F. Manca, S. Giordano, P. L. Palla, R. Zucca, F. Cleri and L. Colombo 2012, J. Chem. Phys, **136**, 154906
[9] H. J. Kreuzer, S. H. Payne, Phys. Rev. E **63**, 021906 (2001)
[10] M. A. Miller, L. M. Amon and W. P. Reinhardt, Chem. Phys. Lett. **331**, 278 (2000).
[11] A. Rosa, T. X. Hoang, D. Marenduzzo and A. Maritan 2005, Biophys. Chem. **115**, 251

Journal of Physics: Conference Series **383** (2012) 012017 doi:10.1088/1742-6596/383/1/012017

Oximetry: a new non-invasive method to detect metabolic effects induced by a local application of mechanical vibration

A. Felici [1,2], C. Trombetta [1,2], P. Abundo [1], C. Foti [3] and N. Rosato [4]

[1] Medical Engineering Service, Tor Vergata University Hospital, Viale Oxford 81, 00133 Rome, Italy

[2] PhD School in Advanced Technology in Rehabilitation Medicine, Tor Vergata University, Via Montpellier 1, 00133 Rome, Italy

[3] Physical and Rehabilitation Medicine Chair, Public Health Dept., Tor Vergata University, Via Montpellier 1, 00133 Rome, Italy

[4] Experimental Medicine Dept., Tor Vergata University, Via Montpellier 1, 00133 Rome, Italy

E-mail: antonella.felici@yahoo.it; chiara.trombetta@yahoo.it; paolo.abundo@gmail.com; foti@med.uniroma2.it; nicola.rosato@uniroma2.it

Abstract. Mechanical vibrations application is increasingly common in clinical practice due to the effectiveness induced by these stimuli on the human body. Local vibration (LV) application allows to apply and act only where needed, focusing the treatment on the selected body segment. An experimental device for LV application was used to generate the vibrations. The aim of this study was to detect and analyze the metabolic effects induced by LV on the brachial bicep muscle by means of an oximeter. This device monitors tissue and muscle oxygenation using NIRS (Near Infrared Spectroscopy) and is able to determine the concentration of haemoglobin and oxygen saturation in the tissue. In a preliminary stage we also investigated the effects induced by LV application, by measuring blood pressure, heart rate, oxygen saturation and temperature. These data confirmed that the effects induced by LV application are actually localized. The results of the measurements obtained using the oximeter during the vibration application, have shown a variation of the concentrations. In particular an increase of oxygenate haemoglobin was shown, probably caused by an increased muscle activity and/or a rise in local temperature detected during the application.

1. Introduction

Mechanical vibrations are increasingly used in Physical and Rehabilitation Medicine clinical practice, for the treatment of specific diseases and in sport rehabilitation. Indeed, the application of mechanical vibration can induce adaptive responses on the human body, as it can stimulate both neuromuscular and skeletal systems [1][2][3]. Local vibration (LV) application allows to apply and act only where needed, focusing the treatment on the selected body segment. This is of great importance, as it should be noted that not always the application of vibration is shown to have positive effects on our body [4]. However, reduced periods of exposure and specific oscillation frequencies, don't entail any negative metabolic effect but instead induce positive body adaptations. The effectiveness of these applications is widely described in numerous publications but the etiology, i.e. how the vibration application is able to elicit such responses

Figure 1. LV Device: control panel, handpiece and fastening system.

in the human body, still remains partially unknown. The intention of this work is to investigate the metabolic effects induced by vibration treatment, in particular those produced during a local vibration treatment using an experimental device dedicated to LV. In order to detect these effects, a device for monitoring muscle oxygenation, the oximeter, has been used. This oximeter, using near infrared spectroscopic technique (NIRS, Near InfraRed Spectroscopy), is able to determinate the concentration of total haemoglobin $[THC]$, oxygenate haemoglobin $[HbO_2]$, deoxygenate haemoglobin $[Hb]$ and oxygen saturation (% ox) in the tissues [5]. The aim of the study is to understand the metabolic effect of vibration on the muscle by studying and analyzing, through an oximetric analysis, the changes of haemoglobin concentrations and oxygen saturation, during LV treatment.

2. Materials and Methods

2.1. Local Vibration Device

In this work we decided to study LV application on a single muscle, the right brachial bicep muscle. For the application of local mechanical vibrations we used an experimental device, designed and optimized by the Medical Engineering Service of the Tor Vergata University Hospital in Rome, in collaboration with the Physical and Rehabilitation Medicine Chair of the Tor Vergata University in Rome and the BoscoSystemLab s.r.l. company (Fig. 1). This device has a digital control panel through which specific parameters, such as time and frequency, can be set. The console is connected to a handpiece in which a motor and an eccentric mass are placed. The vibration is produced by the rotation of this mass activated by the motor. The vibration is delivered onto the area of interest through an adapter, designed to exactly adhere to the body area treated. In addition, this device is provided with a fastening system, therefore has the advantage of being applied without the operator's support during treatment [6]. This avoids the transmission of vibrations to the operator, which, in the long run, could be harmful to the operator himself. The vibration treatment protocol used in this study is based on a training scheme developed by the Physical Medicine and Rehabilitation Department of the Tor Vergata University Hospital but has been slightly revised since it was performed on healthy subject. The treatment involved a series of three repetitions of vibration at 35Hz for 30 seconds and a repetition of 35Hz for one minute, with 60 seconds pause between each repetition.

2.2. NIRS and human tissue

The oximeter allows the detection of $[THC]$, $[HbO_2]$ $[Hb]$ and % ox in the human tissues using NIRS, a spectroscopic method that works in the near infrared region of the electromagnetic spectrum. This device is based on the theory of photon migration through highly scattering media, such as the human tissue, and works injecting near-infrared light into the tissue at different wavelengths. The propagation of light inside biological tissues can be described in terms of photons' flow [7]. While traveling inside tissues, photons can undergo a number of different processes, such as absorption, elastic scattering, inelastic scattering and fluorescence. The main absorbers of NIR light in blood-perfused tissues are metal complex chromophores, such as haemoglobin, bilirubin, cytochrome and water. Therefore generally oxymeters operate in wavelengths between 700 and 850nm, where the absorption spectra of Hb and HbO_2 are maximally separated and there is minimal overlap with H_2O. An oximeter is made of a light source, a detector and a dispersive element that allows to record the intensity at different wavelengths. Measurements are generally obtained using a photometer that, by means of laser diodes or LED, generates light in the infrared spectrum. The light is emitted from a particular optode, which is connected to the instrument through an optical fiber. The light, scattered by the tissues, is then detected by a second optical fiber and is captured by a photomultiplier or a photodiode. Available devices use reflectance-mode NIRS, in which receiving optodes are placed ipsilateral to the transmitter. These optodes exploit the fact that photons transmitted through a sphere will traverse an elliptical path, in which the mean depth of penetration is proportional to the transmitter and receiver optode separation. Numerous studies have shown that the maximum distance between source and detector is about 3.5-4cm, allowing the photons to penetrate into the biological tissue up to a maximum depth of 5cm. Measurement of tissue oxygen saturation and tissue Hb content is determined by the difference in intensity between the transmitted and received light as described by the Beer-Lambert Law.

2.3. Systemic Analysis

In a preliminary stage we investigated the effects induced by LV application, by measuring blood pressure, heart rate, oxygen saturation and temperature. These parameters were measured in 10 subjects and their variations were examined during the vibration treatment. The vibrating device was placed on the right brachial bicep at 4cm from the inner fold of the elbow flexion. For the simultaneous measurements of pressure, frequency and saturation, a multiparametric monitor for vital signs was used (Mod. CS242 Propaq Welch Allyn). The parameters were recorded at different times: at rest, during each vibration and finally after 2 minutes of rest. For the temperature measurement a digital thermometer (Thermometer Fluke 52 II) provided of two thermocouples was used. These allowed monitoring of temperature changes close to the area subject to vibration. The first probe was placed distally, at 2cm from the inner fold of the elbow flexion, while the second probe was placed at 2cm from the vibrating device. This thermometer allows the measurement and storage of instantaneous temperature values in a defined period and calculates the maximum, the minimum and the mean value in that interval. The protocol used to measure temperature is the following: the instantaneous temperature was measured at rest, before and after the positioning of the device on the arm, to detect the possible temperature changes caused by the presence of the fastening system; during each vibration the thermometer was set in order to provide the minimum, maximum and mean values recorded during the treatment; the instantaneous temperature was measured after 30s of rest time; the instantaneous temperatures were measured at the end of the treatment after the first and second minute of rest and after removing the device.

2.4. Metabolic Analysis

In order to detect the metabolic effects induced by LV on the muscle, a tissue oximeter was used (ISS Model 96208: Two Channel, ISS Inc., Champaign, IL USA). This oximeter is based on the frequency domain resolved spectroscopy with the "multidistance" approach [8]. The device works by injecting near-infrared light into the tissue at two different wavelengths. Light sources are made of 8 laser diodes, four of which emit at 750nm and the other four at 830nm [9]. These are coupled to optical fibers of 400μm inner diameter, which guide the light in the measurement probe that is applied directly to the skin of the patient. The emitted light propagates in the tissue, where it is in part absorbed and in part scattered and backscattered. The latter is then collected by a bundle of optical fibers, of 3mm inside diameter, and sent to a photomultiplier tube. The source and the detector fibers are arranged in the probe so that the instrument can acquire data on the sample surface at four known distances between source and detector with a

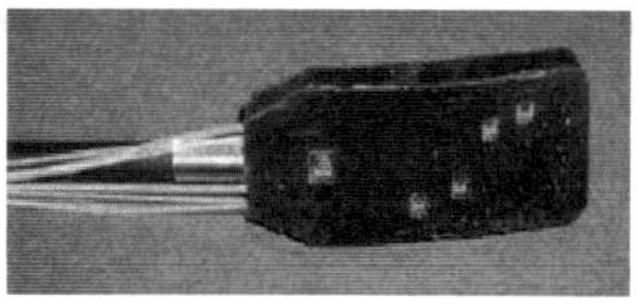

Figure 2. Measurement probe and optical fibers.

constant increase of about 0.5cm, starting from a minimum of 2 to a maximum of 4cm (2, 2.5, 3, and 3.5cm, respectively) (Fig. 2). Light is modulated at a frequency of 110 MHz to allow the collected light phase and modulation measurements. From these raw data the absorption and scattering coefficients of the medium are determined. Once these coefficients are determined, the assumption that haemoglobin is the only significant absorber is applied, so $[HbO_2]$ and $[Hb]$ are calculated [10]. Therefore analyzing these amounts during the treatment, we believed it was possible to obtain information about the metabolic activity of the muscle. For this purpose, in order to observe the presence or absence of these concentrations' variations, the oximeter probe was placed in proximity of the muscle exposed to LV. The measurements were carried out in 5 different subjects. To record parameters, a sampling of 1s rate was performed. In this case, the measurement protocol was modified: four vibrations were applied at 35Hz, lasting 30s for the first three repetitions and 60s for the fourth. Each rest periods was longer than 60s in order to allow the concentrations trends to stabilize at a constant value.

3. Results and Discussion

The data obtained from measurements of pressure, frequency and oxygen saturation, were analyzed using a computation software. From these results it was inferred that while the saturation remained roughly constant during the entire proceeding, the blood pressure and the heart rate values changed but not significantly (Fig. 3). These data confirmed that the effects induced by LV application are actually localized. For what concerns the temperature an increase of the values was noted, that in some subjects reached a maximum of $3°C$. On average, however, the increase detected by the distal probe T1 was $1.5°C$, while the proximal probe, T2, increase was $2.5°C$. From the analysis performed by the oximeter, on the data obtained in each subject and recorded every second, after performing a regression with Matlab program, it was possible to observe a definite trend for each parameter. Fig. 4 shows these trends and their interpolations obtained with the measurements in one of the five examined subjects. In this figure it can be observed how $[THC]$ and $[HbO_2]$ have a very similar trend, particularly after the first two LV applications. These concentrations increased up to a maximum value and then remained constant. Less evident is the increase of the % ox and the $[Hb]$ decrease. The concentration values growth of the first three parameters, in all 5 subjects, was confirmed with a statistical analysis, the one-tailed t-test, with a significance level of 0.05. This test allowed to compare the normalized difference between the maximum value assumed by each parameter and its initial value in each subject, with a baseline vector. This test confirmed the statistical significant increase of $[THC]$, $[HbO_2]$ and % ox, ($p < 0.05$), as observed on the trend of regression, finding a high probability of growth. As regard to $[Hb]$, a two-tailed t-test with 0.05 significance level was used. A statistical significant variation of $[Hb]$ with respect to a baseline vector has been demonstrated ($p < 0.05$).

4. Conclusion

The analysis performed on the pressure, frequency and saturation values recorded during vibration treatment, showed that, although a small variation of these parameters is present, at the end of the treatment similar initials values are reached. Therefore absence of a systemic

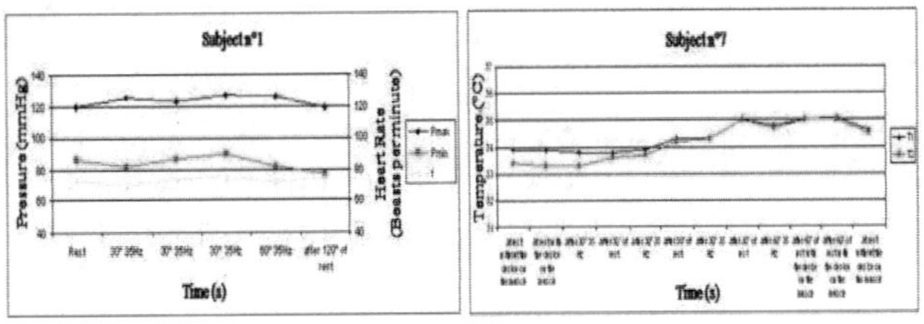

Figure 3. Pressure and heart rate trends related to subject 1; Temperature trends of the two probes related to the subject 7.

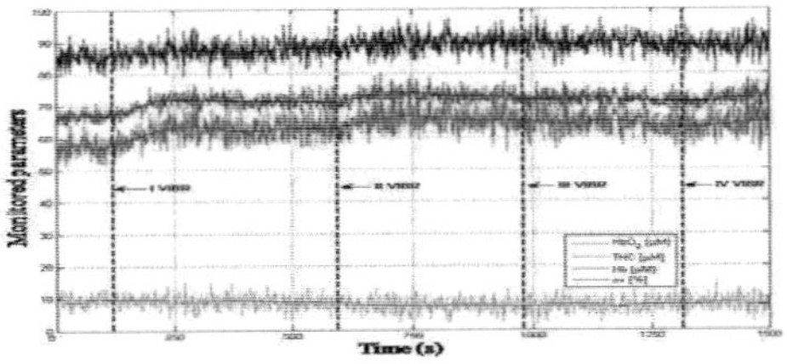

Figure 4. $[THC]$, $[HbO_2]$, $[Hb]$ and $\%\ ox$ during LV treatment.

response induced by the LV application can be affirmed. Moreover, through this preliminary analysis we showed that LV can induce a localized effect leading to an increase of temperature in the treated area. A tissue oximeter was used during LV application to determine haemoglobin concentrations in the tissue analyzed. Through NIRS it was possible to confirm the presence of muscle metabolic activity and identify the concentration variations of the two forms of haemoglobin. This monitoring has allowed to identify in real time the presence of metabolic changes in the muscles during the application of LV. In particular, the trends have shown an increase in oxygenated haemoglobin, that can be justified by an increased blood flow caused by vasodilatation. This could be caused by a muscle activity increase and/or by a local temperature raise, actually encountered during the application of vibration. The next step would be to evaluate if the local temperature increase could have influenced the monitoring carried out with the tissue oximeter. Hence, haemoglobin concentrations could be measured, using the same oximeter, in absence of vibration, inducing an artificial temperature rise in the brachial biceps muscle, equal to the temperature change detected.

References

[1] Trombetta C, Abundo P, Rosato N and Foti C 2011 Application of local vibrations in delayed and non-union fractures: case study. *J. Phys.* Conf. Series **280** (1)

[2] Ljoka C, Della Bella G, Carcelli L, Maugeri E, Giordani L and Foti C 2006 Preliminary study on the effect of the vibratory therapeutic exercise of non-union fracture. *Europa Medicophysica*, **42** (3)

[3] Goodship AE, Lawes TJ and Rubin CT 2009 Low-magnitude high-frequency mechanical signals accelerate and augment endochondral bone repair: preliminary evidence of efficacy. *Journal of Orthopaedic Research*

[4] Abundo P, Trombetta C, Foti C and Rosato N 2011 Production, delivery and application of vibration energy in healthcare. *J. Phys*, Conf. Series **280** (1)

[5] Ferrari M, Mottola L and Quaresima V 2004 Principles, techniques, and limitations of near infrared spectroscopy. *Can. J. Appl. Physiol.*, **29** (4) pp 463–487

[6] Trombetta C, Abundo P, Felici A, Foti C and Rosato N 2011 Development of a device for local vibration application in non-union fractures. *Conference Proceedings, IEEE Interactive Electronic Library* pp 593–596
[7] Taroni P, Pifferi A, Torricelli A, Comelli D and Cubeddu R 2003 In vivo absorption and scattering spectroscopy of biological tissues *Photochem. Photobiol. Sci.* **2** pp 124–9
[8] Fantini S, Franceschini M A, Mayer J S, Walker S A, Barbieri B and Gratton E 1995 Frequency- domain multichannel optical detector for non-invasive tissue spectroscopy and oximetry *Opt. Eng.* **34** (1) pp 32–42
[9] Delpy D T and Cope M 1997 Quantification in tissue near-infrared spectroscopy *Phil. Trans. R. Soc. Lond. B* **352** pp 649–59
[10] Vernieri F, Tibuzzi F, Pasqualetti P, Rosato N, Passarelli F, Rossini P M and Silvestrini M 2004 Transcranial Doppler and Near-Infrared Spectroscopy Can Evaluate the Hemodynamic Effect of Carotid Artery Occlusion *Stroke* **35** pp 64–70

Young Researcher Meeting in Rome 2012 IOP Publishing
Journal of Physics: Conference Series **383** (2012) 012018 doi:10.1088/1742-6596/383/1/012018

Computer Aided Measurement Laser (CAML): technique to quantify post-mastectomy lymphoedema

Chiara Trombetta [1,2], Paolo Abundo [1], Antonella Felici [1,2], Concetta Ljoka[3], Sandro Di Cori[4], Nicola Rosato [5] and Calogero Foti [3]

[1] Medical Engineering Service, Tor Vergata University Hospital, Viale Oxford 81, 00133 Rome, Italy
[2] PhD School in Advanced Technology in Rehabilitation Medicine, Tor Vergata University, Via Montpellier 1, 00133 Rome, Italy
[3] Physical and Rehabilitation Medicine Chair, Public Health Dept., Tor Vergata University, Via Montpellier 1, 00133 Rome, Italy
[4] Reha Group O&P Center, Rome, Italy
[5] Experimental Medicine Dept., Tor Vergata University, Via Montpellier 1, 00133 Rome, Italy

E-mail: chiara.trombetta@yahoo.it; paolo.abundo@gmail.com; antonella.felici@yahoo.it; nicola.rosato@uniroma2.it; foti@med.uniroma2.it

Abstract. Lymphoedema can be a side effect of cancer treatment. Eventhough several methods for assessing lymphoedema are used in clinical practice, an objective quantification of lymphoedema has been problematic. The aim of the study was to determine the objectivity, reliability and repeatability of the computer aided measurement laser (CAML) technique. CAML technique is based on computer aided design (CAD) methods and requires an infrared laser scanner. Measurements are scanned and the information describing size and shape of the limb allows to design the model by using the CAD software. The objectivity and repeatability was established in the beginning using a phantom. Consequently a group of subjects presenting post-breast cancer lymphoedema was evaluated using as a control the contralateral limb. Results confirmed that in clinical settings CAML technique is easy to perform, rapid and provides meaningful data for assessing lymphoedema. Future research will include a comparison of upper limb CAML technique between healthy subjects and patients with known lymphoedema.

1. Introduction

Lymphoedema is a condition of localized fluid retention and tissue swelling due to the accumulation of high molecular weight protein fluid in the interstitial spaces of the affected body part. It is caused by a compromised lymphatic system and can be a side effect of cancer treatment. Often, following breast cancer surgery, for the purpose of local radicality and cancer staging, an axillary lymphadenectomy is performed. The removal of axillary lymph nodes is a predisposing factor to ipsilateral upper limb lymphoedema. However other factors, such as local radiation therapy or cancer progression, may be responsible. Lymphoedema is not the swelling which may occur immediately after surgery and may be present at the post-operation visit, it is a chronic condition that, if ignored, can lead to immobilization and severe infections [1]. These complications can worsen the physical and psychological strain of dealing with breast cancer and significantly impact the quality of life of the patients [2]. Even though several methods for assessing lymphoedema can be used in clinical practice and are reported in literature, an objective quantification, in particular for volume measurements, has been challenging [3] [4]. Despite the fact that various measurement approaches and criteria have been applied, there is no gold standard for measuring or detecting lymphoedema, therefore assessment and diagnosis has been problematic [5]. Most of the measurements are detected on the most affected part of the limb and compared with a control [6]. Finding a 2cm difference or more in limb girth between the affected and non-affected limbs usually guarantees a clinical diagnosis of lymphoedema. Other documented methods of diagnosis measure a 200ml limb volume difference or a 10% limb volume change from baseline and/or between limbs as well as self-reported signs and symptoms [7] [8]. Although these methods can provide data, they also can be time consuming, complicated, expensive or inaccurate. Increased measurement accuracy can improve lymphoedema incidence and prevalence understanding and assist treatment intervention decisions [9].

In clinical practice circumferential measurements are mostly used because they are easy to perform and allow for use of the contralateral limb as a control [10]. However, its objective quantification, especially volume, has always been tricky [11]. The Computer Aided Measurement Laser (CAML), a 3D laser scanning technology, suggests a more sensitive and accurate method that could provide a fast, precise and non invasive technique to quantify upper arm lymphoedema. This technique is based on infrared laser scanning and computer aided design (CAD). Measurements are collected by the laser that scans the object in real-time and then used to design a 3D model of the item using CAD. The aim of this study is to determine the objectivity, reliability and repeatability of the CAML technique in the diagnosis and assessment of upper arm post-breast cancer lymphoedema.

2. Materials and Methods

CAML technique requires an infrared laser scanner and a CAD system. The 3D laser scanner analyzes a real-world object and collects data on its size, shape and appearance. This data can then be used to design a digital 3D model of the tested limb using CAD. The non-contact active scanner, in order to probe an object, emits light and detects its reflection. Laser scanners can send trillions of light photons toward an object and only receive a small percentage of those photons back via optics. Laser flow rate depends on the type of material struck by the beam. Concerning the laser wavelength, the reflectivity of the material changes: greater the reflectivity, greater the flow rate, while lower the reflectivity, therefore greater the proportion of energy absorbed, lower the flow rate. While white surfaces reflect light, black surfaces reflect only a small amount and transparent objects refract the light, therefore laser scanners collect information about surfaces that are not obscured. Optical technologies, however, encounter many difficulties with shiny, mirroring or transparent objects that can change their reflectivity and give false 3D information. A scanning laser system provides, as a direct result of the measurement session, a set of three-dimensional coordinates. 3D scanners collect distance information about surfaces within their field of view and describe the distance to a surface at each point in a local coordinate system relative to the scanner. These coordinates identify a large number of points (point cloud) that define the surface of the scanned object. Laser scanning systems are fully automatic and are able to acquire a considerable number of points per second, sometimes even in the thousands, allowing to operate on small distances offering high accuracy. The purpose of a 3D scanner is to create a point cloud of geometric samples on the surface of the subject. These points are used to extrapolate the shape of the subject, a process called modelling, with CAD software. The information gathered from scanning defines size and shape of the limb. These data are processed, allowing to create a model through which we can determine circumferential and limb volume measurements. Reliability and repeatability of CAML method was initially determined using phantoms and then evaluated on a group of patients presenting post-breast cancer upper limb lymphoedema. In this study the Physical and Rehabilitation Medicine Chair and the Medical Engineering Service of the Tor Vergata University in Rome in collaboration

Figure 1. FastSCAN Cobra Polhemus handheld laser scanner.

with the Reha Group O&P Center used the FastSCAN Cobra handheld laser scanner built by Polhemus to scan the phantoms and the upper arm of the enrolled patients. The FastSCAN works by projecting a fan of laser light on the object while the camera views the laser to record cross-sectional depth profiles. 3D surfaces are instantly acquired gathering measurements by sweeping a handheld laser scanning wand over an object. The finished scan is processed to combine any overlapping sweeps, significantly reducing the time to develop surface models. A full three-dimensional surface of the object is reconstructed in real-time, without the need to place untidy or unwanted registration marks on the object itself. This 3D data is then exported to a host computer with 3D CAD programs. FastSCAN resolution along the laser line depends on wand-object range, typically 0.5mm to 200mm. Scanning rate is 50 lines/second and line-to-line resolution depends on wand movement, usually 1mm at 50mm/second. Scanning range from wand to transmitter and/or receiver to transmitter can be selected from a radius up to 75cm. Operation in presence of metal objects or electromagnetic fields may interfere with the scanner's tracking and degrade performance. Therefore transmitter, wand and receiver must be kept sufficiently far from metal objects or fields. The FastSCAN system includes a System Electronics Unit, a power supply, a receiver and a transmitter. The Electronics Unit contains the hardware and software necessary to generate and sense the magnetic fields, compute position and orientation, control the handheld laser scanner and the interface with the host computer. The transmitter is a triad of electromagnetic coils, enclosed in a plastic shell, that emits the magnetic fields. Handheld laser scanner contains the laser unit, a video camera which records the three-dimensional surface of the scanned object, and the receiver, which detects the position of the scanner. The reliability and repeatability of the CAML method were initially determined using two truncated cone-shaped phantoms, made using a lathe. These two cones of 34cm height and 14.5cm small diameter, differ having a greater diameter of 20.5cm and 35cm (Fig.2). 53 scans were performed from which, using the CAD software, it was possible to determine the volumes of the cones and the circumferences' values of the upper and lower surfaces of the two cones. These were compared with the cone values measured, to determine the accuracy. In addition we analyzed the possible covariates, such as the different operators who performed the scan and the different days of measurements, to confirm the objectivity of the measure. In a second phase of the project CAML technique was evaluated on a group of patients presenting post-mastectomy upper limb lymphoedema, using the contralateral limb as control. In this analysis the actual presence of lymphoedema has been verified by comparison with the healthy limb. In order to validate the objectivity, reliability and repeatability of the CAML technique in the diagnosis and evaluation of upper arm post-mastectomy limphoedema, a statistical analysis of data was carried out using calculating and STATA/SE softwares.

3. Results

For what concerns the phantoms, reliability was measured and confirmed comparing the scanned measurement values with the numerical values calculated. The normalized data obtained were analyzed using the one-tailed t-test with a 0.05 alpha value, assuming an acceptable error value

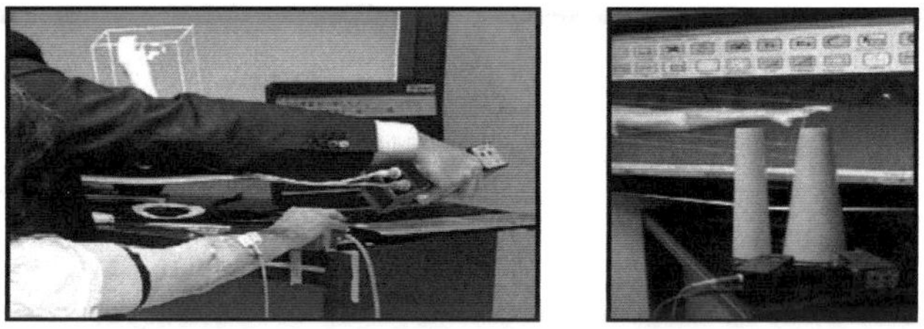

Figure 2. Phantom used to determine measurements' repeatability and reliability and image of the scanned limb model obtained with CAML technique.

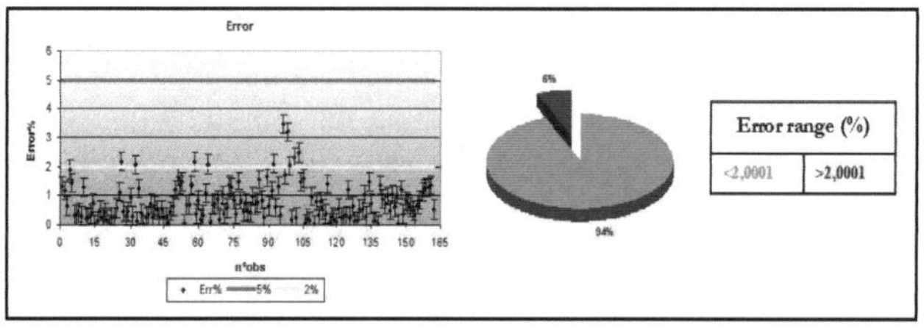

Figure 3. Reliability results.

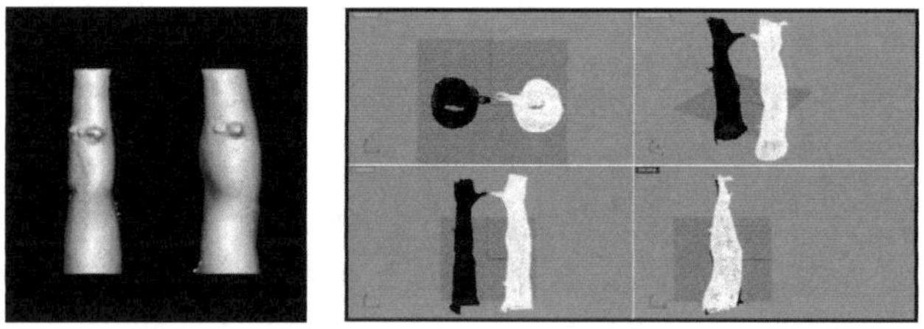

Figure 4. 3D Model and Mesh of the limb affected by lymphoedema and its contralateral obtained with CAML technique.

of 2%(Fig. 3). 94% of the normalized data, indicating the measurement error, were found to be less than the chosen error value. Therefore the method is able to provide reliable measures of circumference and volume.
Regarding the determination of repeatability, circumference and volume measures of the phantoms were analyzed with the ANOVA test, as a function of the two covariates, operator and time. For what concerns the circumferential analysis only time was used as covariate, while in the volume analysis both covariates were studied. A Oneway ANOVA test with Bonferroni correction was performed for the covariate time. Results showed a value of $F = 1.55$ for the circumferences and a value of $F = 2.41$ for volumes. For volume data both covariates, operator and time, were considered in the twoway ANOVA test. Results determined a value of $F = 0.98$. By comparing these F values with the F critical values from the known statistical tables, a statistical significance wasn't obtained. This confirmed that the mean values between groups are the same and therefore the repeatability of the measure is established. A statistical analysis of the data obtained from the upper limb measurements of patients with post-mastectomy lymphoedema was also performed (Fig. 4). In this case the volume values of the limb affected by lymphoedema were compared with those of the healthy contralateral limb by using the one-tail t test. A significant statistical difference was shown ($p = 0.029$), confirming an increased volume of the limb presenting lymphoedema.

4. Conclusions

Upper limb lymphoedema is a distressing sequel of breast cancer treatment. If upper limb volume discrepancies are detected in early post-mastectomy lymphoedema, treatment may be instituted quickly, leading perhaps to a lesser impact on the patient's quality of life and body image. In order to gauge the efficacy of upper limb lymphoedema treatment, limb circumference and volume need to be measured. Accurate, valid, reliable, specific, sensitive and practical tools are required to measure upper limb volumes. CAML technique provides meaningful circumferential and volume data to evaluate post-breast cancer upper limb lymphoedema through a quick and easy implementation. The CAML method is able to determine circumference measures, comparable to data obtained with circumferential measurements, but with minimal margins of error. Above all CAML provides measures of volume, parameter that allows objective quantification of lymphoedema and that is usually difficult to calculate. Future research will aim to study this technique by comparing upper limb CAML data between healthy subjects and patients with known lymphoedema.

References

[1] Armer J M, Stewart B R and Shook R, 2009 Occurrence of Lymphedema Continues To Increase Twelve to Thirty Months after Breast Cancer Treatment, *Cancer Res* **69** (24) Suppl Abstract nr 2070

[2] Cormier J N, Xing Y, Zaniletti I, Askew R L, Stewart B R and Armer J M, 2009 Minimal limb volume change has a significant impact on breast cancer survivors, *Lymphology* **42** (4) pp 161–75

[3] Sander A P, Hajer N M, Hemenway K, Miller A C, 2002 Upper-extremity volume measurements in women with lymphedema: a comparison of measurements obtained via water displacement with geometrically determined volume *Phys Ther* **82** pp 1202–12

[4] Ridner S H, Montgomery L D, Armer J M and Stewart B R, 2007 Comparison of upper limb volume measurement techniques between healthy volunteers and individuals with known lymphoedema *Lymphology* **40** 1 pp 35–46

[5] Armer J M and Stewart B R, 2005 A comparison of four diagnostic criteria for lymphoedema in a post-breast cancer population *Lymphat Res Biol* **3** (4) pp 208–17

[6] Taylor R, Jayasinghe U W, Koelmeyer L, Ung O, Boyages J, 2006 Reliability and validity of arm volume measurements for assessment of lymphedema *Phys Ther* **86** (2), pp 205–14

[7] Armer J M, Stewart B R and Shook R, 2009 Breast cancer survivors often meet multiple criteria for lymphoedema *Cancer Res* **69** (24) Suppl Abstract nr 2074

[8] Armer J M, Stewart B R and Shook R, 2009 30-Month Post-Breast Cancer Treatment Lymphoedema *J Lymphoedema* **4** (1) pp 14–18

[9] Armer J M, 2005, The problem of post-breast cancer lymphedema: impact and measurement issues *Cancer Investigation* **23** (1) pp 76–83

[10] Brown J A, 2004, Clinically useful method for evaluating lymphoedema *Clin J Oncol Nurs* **8** (1) pp 35–8

[11] Deltombe T, Jamart J, Recloux S, 2007, Reliability and limits of agreement of circumferential, water displacement and optoelectronicvolumetry in the measurement of upper limb limphedema *Lymphology* **40** pp 26–34

AUTHOR INDEX

Institute of Physics
Dirac House, Temple Back
Bristol BS1 6BE UK